Eleanor B.

Stanford

1 b -

November, 1961

AN INTRODUCTION TO
CRYSTALLOGRAPHY

AN INTRODUCTION TO CRYSTALLOGRAPHY

BY

F. C. PHILLIPS, M.A., Ph.D.

Lecturer in Geology in the University of Bristol

WITH 515 DIAGRAMS

LONGMANS

LONGMANS, GREEN AND CO LTD
6 & 7 CLIFFORD STREET, LONDON W1
THIBAULT HOUSE, THIBAULT SQUARE, CAPE TOWN
605–611 LONSDALE STREET, MELBOURNE C1
443 LOCKHART ROAD, HONG KONG
ACCRA, AUCKLAND, IBADAN
KINGSTON (JAMAICA), KUALA LUMPUR
LAHORE, NAIROBI, SALISBURY (RHODESIA)

LONGMANS, GREEN AND CO INC
119 WEST 40TH STREET, NEW YORK 18

LONGMANS, GREEN AND CO
20 CRANFIELD ROAD, TORONTO 16

ORIENT LONGMANS PRIVATE LTD
CALCUTTA, BOMBAY, MADRAS
DELHI, HYDERABAD, DACCA

FIRST PUBLISHED 1946
NEW IMPRESSION 1948
NEW IMPRESSION 1951
SECOND EDITION 1956
NEW IMPRESSION 1957
NEW IMPRESSION 1960

PRINTED IN GREAT BRITAIN BY ROBERT MACLEHOSE AND CO. LTD
THE UNIVERSITY PRESS, GLASGOW

PREFACE TO THE SECOND EDITION

The welcome accorded to the first edition of this book affords gratifying evidence that it fills efficiently a gap in the literature of crystallography. It has not been necessary, therefore, to make any fundamental changes in this edition. Some minor emendations have been effected in the text and a few figures improved. The chapter on Space Groups has been modified to bring the symbolism to close accord with that now standardised in the new *International Tables for X-ray Crystallography*. In response to many requests I have intro. duced one new chapter, on the Diffraction of X-rays by Crystals. Though this makes no attempt to equip the student to undertake practical work in X-ray crystallography it does enable the *Introduction* to be carried to a more satisfying conclusion by explaining how use is made of the information in chapters X and XI during the study of crystal structures. In deciding on the scope of this new chapter I have received valuable suggestions from Prof. L. G. Berry and Dr. N. F. M. Henry. The final chapter of the first edition, on Crystal Habit, received a mixed reception by critics; whilst some considered it out of place in an elementary work, others welcomed it as an acceptable (even if unexpected) introduction to an aspect of crystallography which is often neglected but of growing importance. In view of the interest which it has aroused I have retained it almost unchanged.

Bristol, 1955   F. C. PHILLIPS

▼

PREFACE TO THE FIRST EDITION

Text-books of science, in the mind of the discerning critic, usually fall readily into one or the other of two groups, the helpful and the impressive, accordingly as the author's outlook is directed mainly towards the reader's progress or towards the enhancement of his own reputation. I cannot claim that this book is anything more than ' un ouvrage d'enseignement ', in which I have tried to set clearly before the student the elements of the science of crystallography. There appears to be a real need for such a text. The early development of crystallography lay almost entirely in the hands of mineralogists, and excellent text-books of mineralogy exist. More recently, the expanded interest in crystallography consequent upon the rapid development of the study of crystal structure has prompted the production of crystallographic texts without any mineralogical emphasis. Many of these are small books designed to interest those who wish to learn something of the achievements in this field without themselves embarking on crystallographic studies; most of the larger volumes are written primarily for physicists, and aim at imparting just sufficient knowledge of elementary crystallography to allow the student to pass on quickly to the application of X-ray methods to the study of internal structure.

The fact that the main centre of interest in crystallographic studies has been changed by the discovery, by Friedrich, Knipping and von Laue, of the diffraction of X-rays by crystals is indisputable. As a consequence, the belief is now widely held that external morphology is no longer of interest or importance, and we are urged to adopt a ' new view-point ' and to begin the study of crystallography in terms of the structural pattern of crystals. Twenty years' experience of teaching the subject, however, has convinced me that an historical approach is still by far the best for elementary students. The critic will look in vain through this *Introduction* for any detailed exposition of the interaction of X-rays and crystals, not because I am inappreciative of the immense importance of recent achievements in this field but because I hold firmly that what I have presented here is the minimum of basic knowledge essential to real progress in any branch of crystallography.

It is not the least serious drawback of teaching from the ' new view-point ' of the conception of a pattern based on a space lattice that the student is asked to accept at the outset so much that he cannot immediately investigate for himself. He cannot see and handle the atomic structure, and check for himself the regular arrangement, in the same direct way in which he can handle the crystals themselves and check the regularity of the angular relationships of the faces by direct goniometrical measurements until the existence of an orderly structure in the crystalline state becomes something much more real to him than a plausible explanation of certain diffraction effects. A friendly critic has suggested that I should describe this book as an introduction to *classical* crystallography, but I am convinced that it can fairly be considered an introduction for all who hope one day to claim the title of crystallographer.

The illustrations throughout have been specially drawn. A preliminary review of published figures revealed so many mistakes in standard reference works that this seemed to be the only safe course. Though real progress can be made only by handling actual crystals and crystal models, it is essential to any understanding of a book of this kind that it should be freely illustrated. I am greatly indebted to the publishers and to their draughtsman, Mr. H. C. Waddams of Emery Walker Ltd., for the care which they have devoted to the preparation of my original drawings for reproduction, though I cannot hope that I myself have avoided all errors. The figure of a stereographic net is reduced, by permission of the Council of the Mineralogical Society, from a net of $2\frac{1}{2}$ ins. radius originally published by the late Prof. A. Hutchinson in the Mineralogical Magazine.

In the chapter on mathematical relationships I have tried to be reasonably exact without becoming ponderous, keeping in mind the needs of the student of limited mathematical ability. Mathematically-minded readers can derive proofs where I have omitted these. It seemed essential to present a proof of the fundamental Law of Rational Sine Ratios, and I have chosen one combining simplicity with reasonable elegance ; though often ascribed to G. Cesàro, who published a version of it in 1916, it is essentially the same as one used much earlier by Story-Maskelyne.

The development of the thirty-two crystal classes and the discussion of space groups are conducted in the Hermann-Mauguin notation. The only manageable notation for space groups, this is certainly also the most elegant for the crystal classes, and it is greatly to be hoped that it will be adopted also by those whose primary concern does not pass beyond external morphology. In view of its use in the authoritative *International Tables for the Determination of Crystal Structures* I have accepted this notation almost without modification, though on a few points I have ventured to express a personal opinion. Chapter XI is not to be regarded as a rigid derivation of all the space groups but rather as an indication of the manner in which a more formal mathematical approach enabled these groups to be built up. It is essential that the student should be trained from the outset to picture a space group as a three-dimensional scaffolding of symmetry elements, and I have therefore introduced clinographic views of certain groups though such figures are liable to be confusing in all but the simplest examples.

The concluding chapter deals briefly with a subject which has been strangely neglected by British crystallographers; I hope that it will support in the mind of the reader my contention that the study of crystal habit is still far from being a matter of mere historical interest.

As a teacher I owe much to the generations of students who have passed through my classes and to my colleagues, past and present, who have given me generous help. In particular, I am indebted to Dr. N. F. M. Henry and Dr. W. A. Wooster for much constructive criticism of the manuscript and to Mr. A. G. Brighton and Dr. Henry for invaluable help in the correction of proofs.

'Le but de l'enseignement, et surtout celui de l'enseignement supériéur,' wrote Friedel, ' doit être moins d'instruire que d'éduquer et de faire réfléchir; moins d'entasser des connaissances que d'apprendre à en digérer quelques-unes; moins de glisser sur les difficultés que de les mettre en lumière; moins de laisser croire à l'infaillibilité des méthodes en usage et à la certitude des résultats que d'en montrer les points faibles et de cultiver ainsi l'esprit de critique et de libre examen, base nécessaire de l'esprit de recherche.' This book is an introduction; its success will be measured by the number of its readers who finally lay it aside and, ' throwing off the shackles of the text-book ', set out upon their own crystallographic investigations.

Cambridge, 1946   F. COLES PHILLIPS

CONTENTS

PART I
THE EXTERNAL SYMMETRY OF CRYSTALS

CHAPTER I
THE NATURE OF THE CRYSTALLINE STATE

Crystallography is the science of crystals, and so we must ask ourselves at the outset—what is a crystal? To most of us, the word recalls at once such familiar examples as sugar, salt or ice, or the alum which we grew from aqueous solution in early experiments in the school laboratory. We thus fasten immediately upon some of the essential characteristics—a crystal is a solid bounded by a series of plane ' faces ' and usually these faces appear obviously to have some kind of regularity of arrangement. We may be familiar with the beautiful examples of crystals displayed in a mineralogical museum, and the word κρύσταλλος (ice) was indeed first applied to the naturally-occurring oxide of silicon, the common mineral quartz, which was thought to be water congealed by intense cold. From the mineral specimens of the museum it is a natural step to the faceted gems mounted in rings and other articles of human adornment, but the step may lead us into a popular error. A cut gem is indeed bounded by plane faces regularly disposed, but their disposition is at the whim of the lapidary who cut and polished the stone, and is determined largely by the size and shape of the particular specimen on which he is at work. True crystal faces, on the other hand, are the outcome of a natural process, natural in the sense that, even if crystallisation is taking place under controlled conditions in a laboratory, the nature and disposition of these faces are directly related to the process of growth of the crystal without human interference. We can define a crystal as a homogeneous solid bounded by naturally-formed plane faces. The arrangement of these faces is an expression of the manner in which the matter of the crystal is assembled as it grows, and we shall find abundant evidence in the course of our work that this assemblage takes place in a regular manner, so that the naturally-formed external faces which we study are related to a *regular internal arrangement*.

During the earlier part of our investigations we shall be engaged in attempting to draw from a study of these external faces inferences about the nature of this internal arrangement, until it ultimately becomes clear that it is the pattern of this arrangement which is all-important. Not only the external shapes of its crystals, but all the physical properties of a particular substance, depend upon its particular internal structure. After a close study of all the properties of crystals of corundum, Al_2O_3, for example, we shall be able to dispense with the external shape as an aid to identification; a fragment of such a crystal, bounded externally only by irregular fractures, still possesses the same internal structure resulting, for instance, in the same optical properties. If we are unfortunate in our choice of jeweller, and have been sold a ' paste ' (glass) imitation of the gem we are seeking, the artificial nature of the external facets will no longer be a bar to discovering the deception, for such a glass lacks the regular internal structure—it is not *crystalline*. Finally, after a thorough study of the orderly arrangements underlying the well-formed crystals which will be our main subject of discussion here, we shall be in a position to recognise less complete orderliness, until eventually we come to discover some degree of crystallinity in many substances not capable of existing in well-formed single crystals. The modern crystallographer includes in his field of study such initially unpromising materials as rubber and synthetic plastics, silk and wool, and by the same methods investigates even liquids and gases. Crystallography is no longer merely the science of crystals, but of the *crystalline state*. If we confine our attention here mainly to the narrower field of well-developed crystals, it is only to lay a secure foundation on which to erect the towering superstructure in which modern crystallography, in the widest sense, has found some of its most striking applications.

CRYSTAL SYMMETRY

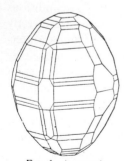

FIG. 1. A crystal of sulphur.

One of the most noticeable features of many crystals, as we have already seen, is a certain regularity of arrangement of the faces, and we proceed to study the nature of this regularity in greater detail.

If we examine a number of drawings of typical crystals or, better still, a number of crystal models, it is at once apparent that there is a strong tendency for faces to be so arranged that the edges formed by a number of them are parallel. This feature is very evident, for example, in the crystal of sulphur represented in Fig. 1. Such a set of faces constitutes a *zone*, which we can define as a set of faces whose mutual intersections are all parallel. The common direction of edge is that of the *zone axis* of that particular zone.

The next regular feature we might notice is the frequent occurrence of similar faces (of the same size and shape) in parallel pairs on opposite sides of the crystal. Many crystals are bounded entirely by such pairs of faces, the sulphur crystal of Fig. 1, for example, and are said to show a *centre of symmetry*. A solid such as the regular tetrahedron, however, in which a face on one side is opposite a point (or *coign*) on the other, does not possess a centre of symmetry. In examining more complex models, we shall sometimes discover examples in which some of the faces occur in parallel pairs whilst others have no similar face parallel to them; a crystal does not show a centre of symmetry unless every face has a similar face parallel to it.

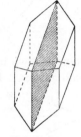

FIG. 2. A crystal showing a plane of symmetry.

Many of the models will show another kind of regularity of arrangement; the crystal in Fig. 2 is bilaterally symmetrical—it shows a *plane of symmetry*. Highly regular crystals may be bilaterally symmetrical about several planes cutting them in different directions; they have several planes of symmetry. Thus in a cube there are three planes of symmetry of the kind shown in Fig. 3, parallel to the faces of the

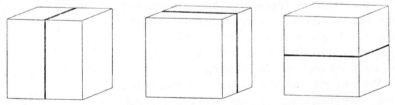

FIG. 3. The three planes of symmetry parallel to the faces of a cube.

cube; but there are also six diagonal planes, shown in Fig. 4. There is one important characteristic, however, about a *crystallographic plane of symmetry* which differentiates it from our ordinary conception of a plane of geometrical symmetry; not only must the plane be such that it divides the crystal into two equal portions, but these two portions must be so situated that they are mirror images of each other with respect to the plane. Thus, while a cube has the six diagonal symmetry

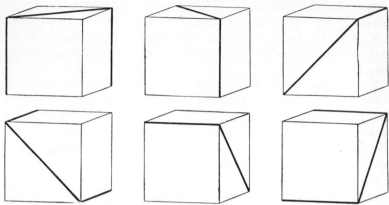

FIG. 4. The six diagonal planes of symmetry in a cube.

planes shown in Fig. 4, the rectangular parallelepiped of Fig. 5 has no such planes.   The plane marked in this figure does divide the solid

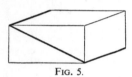

FIG. 5.

into two geometrically similar wedges, but they are not situated as reflections of each other in the plane.   The only crystallographic symmetry planes present here are parallel to the three pairs of faces.

A third kind of crystallographic symmetry is symmetry about a line, termed an *axis of symmetry*.   If a cube is rotated about a line normal to one of its faces at its mid-point (Fig. 6), it will turn into a congruent position every 90°, and therefore four times during a complete revolution;   the normal is an axis of fourfold symmetry, a *tetrad* axis, and a cube clearly possesses three such axes, one normal to each of the three pairs of parallel faces.

We can thus define an axis of symmetry as a line such that after rotation about it through 360°/n the crystal assumes a congruent position; the value of n determines the *degree* of the axis.

If $n = 1$, the crystal must be rotated completely through 360° before congruence is achieved. Such an axis is termed an *identity axis*, and every crystal clearly possesses an infinite number

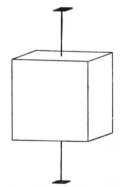

FIG. 6. One of the tetrad axes of a cube.

of such axes.   This concept is of little use to us at present, but will be helpful later in our study.

If $n = 2$, the crystal must be rotated through 180°, and the axis is termed a *diad* axis.

If $n = 3$, congruence is achieved every 120°, and the axis is a *triad*.

If $n = 4$, the corresponding angle of rotation is 90°, and the axis is a *tetrad*.

If $n = 6$, giving congruence every 60°, the axis is a *hexad*.

We shall be in a position later to prove that these are the only possible values; investigation of other laws of crystal architecture will show that a degree of symmetry higher than 6 is impossible, and will also account for the absence of the value $n = 5$ (a pentad axis is not a possible crystal symmetry axis).

We have already seen that a cube possesses a centre of symmetry, nine planes of symmetry (three of one kind, and six of another) and three tetrad axes. It has also other axes of symmetry; it may be rotated about a solid diagonal through 120° to reach congruence (Fig. 7), and such a line, of which there are four, is therefore a triad axis.

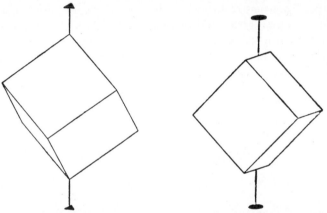

FIG. 7. One of the triad axes of a cube.   FIG. 8. One of the diad axes of a cube.

Finally, a line joining the middle points of a pair of opposite parallel edges proves to be a diad axis (Fig. 8), and there are six of these present in the cube. The full crystallographic symmetry of the cube is thus:

centre of symmetry
3 planes
6 diagonal planes } 9 planes, Figs. 3, 4.
3 tetrad axes
4 triad axes } 13 axes, Fig. 9.
6 diad axes

By handling models, it soon becomes evident that this same group of symmetry elements is present in many other crystals of quite differ-

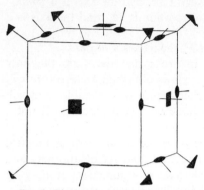

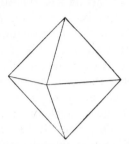

FIG. 9. The thirteen axes of symmetry shown by a cube.

FIG. 10. The octahedron.

ent shapes from that of the cube. It is the symmetry, for example, of the octahedron (Fig. 10) and of the rhombic dodecahedron (Fig. 11). A rhombohedron (Fig. 12), on the other hand, shows considerably less symmetry. It may be looked upon as derived from a cube by compression (or extension) along one of its triad axes. The upper

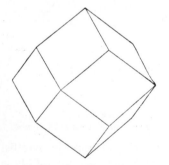

FIG. 11. The rhombic dodecahedron.

FIG. 12. A rhombohedron.

and lower coigns are no longer right-angled, but the diagonal joining them is still a triad axis. It is the only triad axis which the model possesses, however, for the six remaining coigns are formed by two kinds of edge, a *polar* edge *ab* running down from the emergence of the triad axis and two edges forming part of the zig-zag 'waist-line' of the model. There is no axis of symmetry (other than an identity axis) passing through these six coigns, and the full symmetry of the rhombohedron can be worked out to be

Centre of symmetry,
1 triad axis,
3 diad axes,
3 planes.

Just as other models beside the cube show the same characteristic group of symmetry elements which we derived from the cube, so we find the above group, derived from the rhombohedron, shown also by other crystals of quite different shape. A little consideration will show, too, that the total number of different symmetry groups which can be constructed from all the possible kinds of crystallographic symmetry elements is comparatively limited, since the symmetry elements react on each other. A plane of symmetry, for example, will repeat any axis of symmetry inclined to it at an angle other than 90°, so that in the only possible combination consisting of one plane of symmetry and one axis, whether diad, triad, tetrad or hexad, the axis must be normal to the plane. The symmetry of the rhombohedron derived above is an example of a group containing one triad axis; if a second triad axis were present, inclined to the first, there must be three such inclined axes, since by definition the whole group must be rotated into congruence for every rotation of 120° about the first triad. It is thus not possible to have a symmetry group with two triads, or with three triads—if more than one triad is present there must be four, and we have already encountered such a group in discussing the symmetry of the cube.

THE SEVEN CRYSTAL SYSTEMS

On the basis of considerations of this kind, crystals are grouped according to their symmetry into seven major divisions, the seven *Crystal Systems*. We shall define these systems at present in terms of axes of symmetry.

The *Triclinic* (Anorthic) system has no axes of symmetry.

The *Monoclinic* system has one diad axis (and no axes of higher degree).

The *Orthorhombic* system has three diad axes.

The *Tetragonal* system has one tetrad axis.

The *Cubic* (Regular, Isometric) system has four triad axes.

The *Trigonal* system has one triad axis.

The *Hexagonal* system has one hexad axis.

Though these definitions, and indeed the present method of approach to the study of crystal symmetry, are not quite rigid, and we shall eventually modify them slightly, study of a number of crystal models will show that in spite of the simplicity of the classification there is no difficulty in allotting a symmetry group to the appropriate system. The rhombohedron, for example, belongs to the trigonal system, possessing the characteristic feature of one triad axis. The cube, octahedron and rhombic dodecahedron, showing four triad axes, all belong to the cubic system. A combination of axes not mentioned in the definitions of the systems will be found always associated with some other characteristic combination; any crystal showing three tetrad axes, for instance, will also be found to possess four triad axes, and so falls naturally into the cubic system.

Each of these systems will later be subdivided into a number of symmetry groups, the *Crystal Classes*, all possessing in common the characteristic symmetry of the system. Thus a crystal in the trigonal system may possess a triad axis only as its sole element of symmetry, or a triad axis and a centre, or various other combinations of one triad axis with diad axes or with symmetry planes, or with both. A rigid discussion will reveal that there are in all 32 crystal classes. For the time being it will not be necessary to consider any others than the most symmetrical (*holosymmetric*) class within each system, but we may note here that in some countries it is customary to consider the trigonal system as a subdivision of a larger hexagonal system. There would thus be only six different crystal systems, but the total number of different symmetry groups, the crystal classes, distributed among the systems is of course the same in either arrangement.

FORM AND HABIT

So far, the examples which we have used in the discussion of crystal symmetry have all been composed entirely of similar faces—the six faces of a cube are all equal squares, the twelve faces of the rhombic dodecahedron are all equal rhombuses. Frequently, however, a crystal shows faces of several different shapes. A cube-like crystal may have small equilateral triangular faces developed in place of each coign of the simple cube (Fig. 13), and the full

Fig. 13. A cube with modified coigns.

symmetry of the cube implies that if one coign is replaced by such a triangular face, then all eight coigns must be similarly modified. The crystal then consists of six octagonal faces, which are part of the original cube faces, and eight of the new equilateral triangular faces; it is said to show faces of two different *forms*. This word is used in a special sense in crystallography, and we must be careful to avoid its use if we wish only to imply a general idea of shape. A rigid definition of a form is ' the assemblage of faces necessitated by the symmetry when one face is given '. The full implication of this definition can only be appreciated gradually, as we proceed to study in turn crystals belonging to the different systems, but some important points can be brought out by a comparison of the cube with the rhombohedron.

All eight coigns of the cube are similar (if they were not, the normals to the cube faces would not be tetrad axes), and the appearance of a small face replacing one coign necessitates, as has been remarked, a similar replacement of all eight coigns; the new crystal is a combination of two forms, the six-faced cube and the eight-faced new form. The eight coigns of a rhombohedron are not all alike, six of them (like the one marked *a* in Fig. 12) being like each other and unlike the two similar obtuse coigns (*b*, Fig. 12). If a new form appears replacing one coign *a*, then all six of these coigns must be replaced, and the new form is six-faced (Fig. 14). The obtuse coigns *b* are not modified by

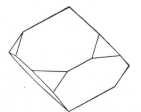

FIG. 14. A rhombohedron with one set of coigns modified.

FIG. 15. A rhombohedron with the trigonal coigns modified.

faces of this new form; they in their turn may be modified by faces of a third form, and if one is so modified they must both be modified or the crystal will no longer possess, for example, a centre of symmetry. This third form therefore consists of two faces (Fig. 15).

Study of crystals composed of more than one form introduces a further new factor. In the single forms, the faces were all the same size and shape; in the cube with its coigns replaced by faces, which we may think of as developed by actually cutting away the original coigns of the cube, what determines how large the new triangular faces should be—how much of the coigns we shall cut off? If only a little is removed, the appearance is that of Fig. 16; if more is cut away, the

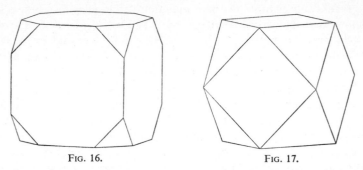

FIG. 16.   FIG. 17.

cube faces may be reduced to squares (Fig. 17), but if the process of paring away is continued further the new faces will meet in a new set of edges and will be hexagonal instead of triangular (Fig. 18). The

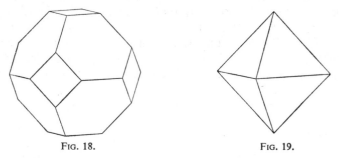

FIG. 18.   FIG. 19.

logical completion of the process is illustrated in Fig. 19, where we are left with only eight faces constituting what is now easily recognisable as the octahedron. Thus all the crystals of Figs. 16, 17 and 18 are cubo-octahedra, combinations of the two forms cube and octahedron, and differ from each other only in the relative development of these forms. Fig. 16 is a cube modified by small faces of the octahedron, whilst Fig. 18 may be described as an octahedron modified by smaller faces of the cube. This relative development is called the *habit*—' the habit of a crystal is the general aspect conferred by the relative development of the different forms '.

Variability of habit is the first source of difficulty on passing from a study of crystal models to a study of actual crystals. It would not be easy, at present, for us to recognise at sight that the crystals of Figs.

16 and 18 are crystallographically identical in the sense that they show the same forms, and differ only in habit. Experiments in the laboratory, crystallising the same substance under different conditions, show that some of the important factors affecting the habit of growth of a particular substance are the conditions of crystallisation—the solvent used, the temperature at which crystallisation takes place, the presence of impurities, and so on. Sodium chloride grows from pure aqueous solution as simple cubes, but the addition of urea to the solution causes crystals to grow resembling Fig. 16, cubes modified by small octahedral faces; potassium chlorate crystallises from pure aqueous solutions as thin platy crystals (a *tabular habit*) with a rhombus-like outline, but the addition of even a trace of certain dyes such as methyl orange changes the habit to that of slender needles (an *acicular habit*). The chemist's customary preparation of his final crystalline product by repeated re-crystallisation from solution in a pure solvent selected from a relatively small range (such as water, alcohol, ether or acetone) results in a reasonably constant habit in a given substance, and he is fortunate that he is not perplexed by the bewildering variety of habit which sometimes confronts the mineralogist, whose crystals have been produced in nature under a wide variety of conditions and from solutions often containing all manner of other substances as impurities.

Models of crystals of varying habit have still one important feature in common; they will all yield on inspection the same group of symmetry elements, and no possible variation of habit can change the crystal class and system to which a given substance is allocated. Whatever the relative sizes of the faces of the two forms in a cubo-octahedron may be, all cubo-octahedra are clearly closely connected, since they all possess the full group of twenty-three elements of symmetry shown by the simple cube. When we transfer our attention to actual crystals we are confronted with a difficulty of a much more acute kind than mere variation of habit. Most of us have grown crystals of alum from solution, and have been told that alum crystallises from aqueous solution in octahedra; though by suspending a small seed-crystal in the solution on a piece of cotton we may finally have succeeded in growing a fairly regular-looking octahedron, most of the crystals formed on the bottom and sides of the beaker probably looked more like Figs. 20 and 21, distorted and irregular-looking, with faces very far from being the same size and shape. Clearly such a crystal would

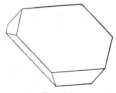

FIG. 20.

not yield on examination many, or even any, of the twenty-three elements of symmetry which we have found in the regular octahedron.

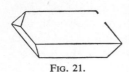

FIG. 21.

How is this difficulty of the irregularity and geometrical distortion of natural crystals to be reconciled with the previous discussion of crystal symmetry conducted in terms of regularly-developed models?

THE LAW OF CONSTANCY OF ANGLE

The acuteness of the difficulty is well brought out by the fact that it held up the discovery of the fundamental law underlying the growth of crystals until as late as 1669. In that year Nicolaus Steno,* a remarkably versatile scientist, published in Latin at Florence a dissertation entitled *De Solido intra Solidum naturaliter contento Dissertationis Prodromus*, a translation of which, ' English'd by H.O.', was published in London in 1671. In this work, amid a variety of geological and mineralogical observations, Steno describes and illustrates measurements which he had made on crystals of the mineral quartz, SiO_2. By cutting sections from differently distorted crystals and tracing their outline on paper he was able to show that analogous angles in the different sections, whatever the actual size and shape of the sections themselves, were always the same. Thus, sections cut at right-angles to the vertical edges *ab* (Fig. 22), though regular hexagons only in an undistorted crystal, always had angles of 120°. Sections cut at right-angles to edges of the kind *ac* gave values different from 120°, but still all such sections could be arranged so that each side of a given section was parallel to a corresponding side of every other section. The right-hand portion of Fig. 22 is a diagrammatic reproduction of one of Steno's figures.

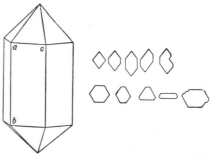

FIG. 22. Diagramatic reproduction of one of Steno's figures, showing sections of a quartz crystal.

* Niels Stensen (Nicolaus Steno) was born in 1638, the son of a goldsmith in Copenhagen where he first studied. His earliest work was in the fields of anatomy and physiology, and at Florence he held the position of physician-in-ordinary to Grand Duke Ferdinand II. He later became interested in geological studies, and the *Prodromus* was planned as a preliminary to a larger treatise which was never published. In 1672 he was appointed Professor of Anatomy in Copenhagen; he died at Schwerin in 1686.

Steno was laying the foundation for the erection of a *Law of Constancy of Angle*, the fundamental law of crystallography. A further century elapsed, however, before the law was firmly established. Steno's work was extended and generalised by Domenico Guglielmini * over the years 1688–1705, and finally confirmed by the work of Romé de l'Isle,† who carried out a very extensive series of measurements and published his results in the period 1772–83. The Law of Constancy of Angle may be formally stated in the following way: ' In all crystals of the same substance, the angles between corresponding faces have a constant value.'

GONIOMETRY

It will be observed that the law refers to *interfacial* angles, and not to the plane angles of the faces themselves. It is these interfacial angles which we must measure when comparing crystals of different development. Steno, as we have described, accomplished this by the rather crude method of cutting sections normal to the edges in question and tracing the outline of such sections on paper. De l'Isle had at first to adopt similar devices, but during the course of their work his assistant, Carangeot, in 1780 invented the earliest type of crystal-measuring instrument, the *contact goniometer*. The essential feature of this device (Fig. 23) consists of two flat bars pivoted together like a pair of scissors and capable of being clamped in any position by means of the screw pivot. The angle between the bars is read off on a graduated semicircle. In simple types such as those used by students when measuring crystal models the straight base of the graduated scale may serve in place of one arm; in others, the two arms can be removed

* Domenico Guglielmini was born in Bologna in 1655. His early work was mainly concerned with hydraulic engineering, but in 1688 he published in Italian at Bologna a paper entitled *Riflessioni filosofiche dedotte dalle figure de' sali*, in which he seems to have accepted Steno's work and to have used it as a foundation for further studies. The paper is remarkable, in particular, as containing one of the earliest suggestions of a theory of crystal structure. Guglielmini was later Professor of Mathematics successively at Bologna and Padua, and his only other publication of crystallographic interest was produced in Latin at Venice in 1705, by which date he had been appointed Professor of Medicine at Padua, where he died in 1710.

† Jean Baptiste Louis Romé Delisle (so spelt on the title-page of his earlier publications; later he wrote de l'Isle) was born in Gray, in eastern France, in 1736, and for the first part of his life followed a military career in the East. Returning to Paris in 1764 on pension, he helped to support himself by giving private lectures in mineralogy. His first publication of interest from our present point of view, *Essai de Cristallographie, ou Description des Figures Géométriques, propres à différens Corps du Règne Minéral, connus vulgairement sous le nom de Cristaux*, appeared in Paris in 1772; and this was followed eleven years later by a much-expanded second edition in four volumes with a slightly modified title. He died in Paris in 1790.

together from the graduated semicircle. The method of use is almost self-evident; the crystal is fitted between the two arms, so that the plane of the instrument is normal to the edge in question, and the two faces are pressed closely one against each arm. The required angle is

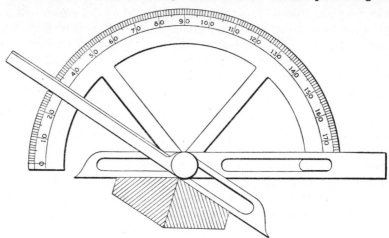

FIG. 23. A contact goniometer.

then read from the graduated scale, the screw pivot being tightened before transference of the arms to the scale when using an instrument with removable arms. In quoting the value of this angle it has become the custom universally (for reasons which we shall shortly appreciate) for crystallographers to use not the actual value of the solid angle but the supplement of this angle, which is the angle between the *normals* to the crystal faces. Thus a crystallographer speaks of a hexagonal prism as having interfacial angles of 60°, and not of 120° as we are at first tempted to say.

Though the contact goniometer is still useful in the examination of large crystals, particularly those with rough or irregular faces, and is the best instrument for use by students in conjunction with crystal models, it has been entirely replaced for all accurate work on small crystals by some type of *reflecting goniometer*. The first description of such an instrument was given by W. H. Wollaston * in 1809. A modern

* William Hyde Wollaston was born at Chislehurst, Kent, in 1766. After graduating at Cambridge he took up medical practice, but relinquished this in 1801 to devote himself to chemistry. His versatility of achievement was remarkable, enabling him to make original discoveries in the fields of pathology, chemistry, crystallography, physics, astronomy and botany. He is well known as the inventor of the *camera lucida* and the discoverer of the elements Palladium and Rhodium. He died in 1828.

version is illustrated in Fig. 24. Suppose that a fixed mirror (Fig. 25) is illuminated by parallel light from a distant source (or from a collimator) and that a part of the beam falls also on the crystal, which is fixed on an axis parallel to the mirror and a short distance above it and so adjusted that the edge over which we wish to measure the interfacial angle is parallel to the axis. The image of a fixed signal such as a horizontal slit is seen, when the eye is suitably placed, reflected in the fixed mirror and also in the upper face of the crystal. The latter is rotated until these two images coincide, a reading

FIG. 24. A reflecting goniometer of the type devised by Wollaston.

of the graduated circle attached to the axis is taken, and the crystal rotated on the axis until coincidence is similarly obtained for the reflected image seen in the second face. This will occur when the second face has been turned into a position parallel to that originally occupied by the first face, and a second reading of the graduated circle gives by

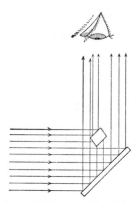

difference the *normal crystallographic angle* between the two faces. Moreover, any further faces belonging to the same zone (p. 3) can obviously be brought into a position to reflect by further rotation, so that after adjusting one pair of faces we can measure with the reflecting goniometer all the interfacial angles in a given zone without further readjustment of the crystal.

In the simplest instruments the axis is horizontal (i.e. the graduated circle is vertical) and the eye is unassisted by a telescope. If there is no collimator (Fig. 24) the signals

FIG. 25. The principle of the Wollaston goniometer.

are provided by a distant lamp or screened window, and it is convenient in practice not to use the image of the same signal reflected both in the mirror and in the crystal face. The former, the 'fixed signal', is usually a horizontal slit, whilst the latter, the 'moving signal', is a small diamond aperture more brilliantly illuminated. The crystal-adjusting apparatus must allow movement parallel to the plane of the graduated circle, for centring the crystal on the axis, and movement in two planes at right-angles, normal to the plane of the circle, for adjustment of the zone-axis parallel to the axis of the instrument.

Greater convenience of manipulation is afforded by further elaboration of the instrument. A collimator provides the signals, and the eye

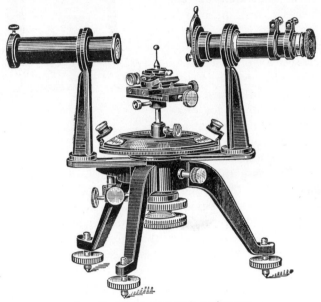

FIG. 26. A horizontal-circle goniometer.

is assisted by a telescope. If the latter is fitted with cross-webs one may dispense with the mirror, and an extra lens swinging in front of the objective enables one to focus the crystal itself for convenience in adjusting. The crystal-adjusting apparatus, too, is improved; two centring screws and two tangent screws provide the necessary movements. Such an instrument is often built as a *horizontal-circle goniometer* (Fig. 26).

In using any single-circle instrument, the crystal must be dismounted and readjusted for measurement of each successive zone. To overcome this disadvantage, more elaborate goniometers have been designed incorporating two, or even three, graduated circles (Fig. 27). Such instruments have special advantages in relationship to particular problems, or for the investigation of very minute crystals. They are not so suitable for the elementary student, because their method of use does not always bring out as clearly as does the single-circle instrument

FIG. 27. A two-circle goniometer.

the all-important zonal relationships. One of the most experienced of crystallographers, Prof. P. Groth,* wrote 'The main problems of crystallographic enquiry can be solved, without any more complicated instruments, by the use of the simple single-circle goniometer.'

Thus furnished with some variety of optical goniometer we are in a position to make more expeditiously the kind of measurements on which Steno and de l'Isle founded the Law of Constancy of Angle. Practical advice on goniometry will be given later (p. 93), but it is already evident in what way we must modify our early discussion of symmetry in order that it may apply to crystals of diverse habit and

* Paul Heinrich Ritter von Groth was born at Magdeburg in 1843. After studying at Dresden, Freiberg and Berlin he was appointed in 1872 Professor of Mineralogy at Strasbourg, moving to the Chair at Munich in 1883. Here he worked for forty years and established for himself an international reputation as a chemical crystallographer. In his five-volume *Chemische Krystallographie* he assembled crystallographic data for over 7000 substances, and in 1877 founded the *Zeitschrift für Krystallographie und Mineralogie*, a periodical in which very many important papers have since appeared. He died at Munich in 1927.

distorted growth—we must seek faces with *similar angular relationships* rather than faces of a given size and shape. Thus, a crystallographic plane of symmetry divides a crystal in such a way that for every face on one side of the plane there is a corresponding face *sloping at the same angle* on the opposite side of the plane; a triad axis produces from a given face two further faces symmetrically disposed around the axis and making the same angle with it, and so for the other elements of crystallographic symmetry.   In fact, *the size and shape of individual faces of a crystal are purely incidental features* determined merely by the conditions of growth of the particular crystal under consideration, but the angular relationships of these faces reveal the underlying crystallographic symmetry.   By measuring a model or a regularly developed crystal, the student can establish that the angle over the edge of a regular octahedron is 70° 32′; measurement of distorted crystals of alum (Figs. 20, 21) reveals that whatever the degree of geometrical distortion the angle over any edge between two adjacent faces still has this same value.

CHAPTER II

METHODS OF PROJECTION

CRYSTAL PROJECTION

Since the sizes and shapes of the faces of a crystal are merely incidental, their variation serves only to obscure the true symmetry relationships, and a discussion of such relationships is carried out most conveniently in terms of some representation of the crystal in which the essential angular relationships are preserved whilst the trivial features are unrepresented. This is achieved by some type of *crystal projection*, in which each face is represented by a dot, which has neither size nor shape but has still particular angular relationships with respect to other dots in the projection.

If from an origin within the crystal we imagine normals to be drawn to all the faces (extended in their own plane if necessary), these normals radiate from the origin in directions depending upon the crystallographic interfacial angles; whether a given face on a particular crystal is large or small its normal will still have the same direction, relative to the normals to other faces, as in every other crystal of the same substance. Illustrating the matter first in two dimensions, Fig. 28

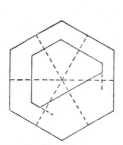

Fig. 28. Normals to a zone of faces at 60°.

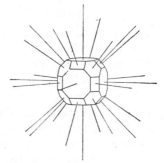

Fig. 29. The bundle of normals to the faces of a crystal.

demonstrates that a zone of faces at 60° will give the same set of normals whatever the dimensional distortion of individual faces. Carrying this process through in three dimensions (Fig. 29) and concentrating our attention only on the bundle of radiating normals, we have in part achieved our objective, but such a bundle is difficult in conception and awkward for pictorial representation, and so the projection is carried further. A sphere is imagined described about the crystal with its centre at the origin of the normals, which are then produced to cut the surface of the sphere, and at each point of intersection a dot (the

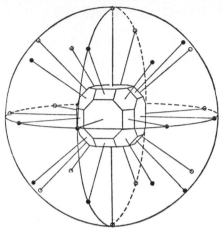

FIG. 30. Spherical projection of the normals of Fig. 29.

pole of the corresponding face) is marked on the sphere (Fig. 30). The essential angular relationships of the crystal are now represented in *spherical projection*.

STEREOGRAPHIC PROJECTION

Such a projection, however, is still three-dimensional, and we finally adopt some device for representing the projection on a plane sheet of paper. The problem is similar to that of the geographer who attempts to represent the surface of the earth on a plane map, and we might choose any one of the various methods which he at times adopts. The particular choice will be dictated by a desire to preserve angular truth, so far as possible, in the projection, whilst areal truth is of little concern. The method most widely used in crystallography is that of the *stereographic projection*, a projection known in ancient Greece in the second century B.C. but first utilised in this way by F. E. Neumann * in 1823 and subsequently brought into general use by W. H. Miller.†

* Franz Ernst Neumann was born at Joachimsthal in 1798, and studied in Berlin. His book *Beiträge zur Krystallonomie* was published in 1823. A few years later he was appointed Professor of Mineralogy and Physics at Königsberg, and his interests gradually turned to the wider field of mathematical physics, in the development of which in Germany he played an important part. He died in 1895.

† William Hallowes Miller was born at Velindre, near Llandovery, S. Wales, in 1801. After graduation at Cambridge as fifth Wrangler, he first occupied himself

The plane of the paper is regarded as passing horizontally through the centre of the spherical projection, which it intersects in the *primitive* circle (Fig. 31). Each pole on the sphere is then projected on to the

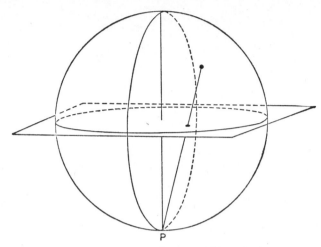

FIG. 31. The principle of stereographic projection.

plane of the paper by joining it to the lowermost point *P* of the sphere, the pole being marked by a small dot on the paper at the point of intersection of this join. The upper half of the crystal thus projects as a series of poles lying within the primitive (Fig. 32), whilst the poles of any faces normal to the paper lie on the primitive itself. If the construction be applied to poles on the lower hemisphere, the joining line must be produced to intersect the paper, and the corresponding pole in projection will lie outside the primitive. Whilst this is sometimes the most convenient method of representation, it extends the projection unduly, since the projections of poles low down on the sphere near *P* will lie at an almost infinite distance from the centre of the primitive. It is customary, therefore, to restrict the projection to the area within the primitive; to accomplish this, poles on the lower hemisphere are

with mathematical work, but in 1832 he was appointed to succeed Whewell in the Chair of Mineralogy, a position which he occupied to the end of his life. In 1839 he published *A Treatise on Crystallography*, a classic work on mathematical crystallography the substance of which was reproduced in more condensed form later as *A Tract on Crystallography* (1863). A mass of original observations on the crystallography of minerals was incorporated in the second edition of *An Elementary Introduction to Mineralogy* by W. Phillips, which Miller produced jointly with H. J. Brooke in 1852. In another field, he did valuable work in connection with the construction of a new Parliamentary standard of weight. He died at Cambridge in 1880.

joined *upwards* to the point P' on the sphere diametrically opposite P, and to indicate that this has been done, and that the pole in projection

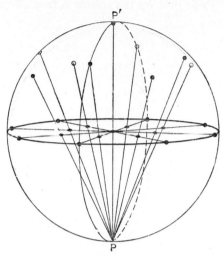

FIG. 32. Stereographic projection of the upper hemisphere of the spherical projection of Fig. 30. The poles marked as open rings are in the rear.

represents a face on the lower half of the crystal, the intersection is marked on the paper as a small open ring instead of a dot.

We now proceed to project stereographically the crystal represented in Fig. 33, showing the faces of the cube, the octahedron and a third

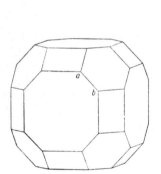

FIG. 33. A modified cube.

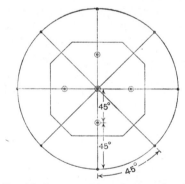

FIG. 34. Stereographic projection of some of the faces of the crystal in Fig. 33.

form with twelve faces. The cube faces can be readily inserted; the pole of the uppermost face lies at the centre of the primitive, and the lowermost face, parallel to this, is represented by a small ring drawn around this dot (Fig. 34). The four vertical faces give poles lying on the primitive itself (notice that dots only will appear on the primitive; a ring in this position could only mean the same face, normal to the paper, which the dot represents).

Considering next the twelve rectangular faces, it is clear from the parallelism of edges (Fig. 33) that they lie in sets of four in zones with the cube faces. Normals to all the faces in one zone will lie in a plane at right-angles to the zone-axis, and this plane will cut the sphere in a *great circle*. (A great circle on a sphere is a section of the surface of the sphere by a plane passing through the centre of the sphere; the Meridians and the Equator are great circles on the terrestrial globe.) In the projection (Fig. 34) one of these great circles is the primitive and the others, vertical great circles passing through the projecting point P, project as diameters of the primitive. Measuring the angle from a cube face on to one of the adjacent rectangular faces gives an interfacial angle of 45°. The four poles on the primitive are readily inserted in the projection, by means of a circular protractor, but for the remaining poles we require to find the correct distance in projection corresponding to an angular distance of 45° from the summit of the sphere (Fig. 34). From a vertical section of the spherical projection (Fig. 35), the distance S, from the centre of the primitive in projection, of the pole of a point at an angular distance θ from the summit of the sphere is seen to be $r \tan \theta/2$, where r is the radius of the sphere of projection (and therefore of the primitive). Using a value $\theta = 45°$ this distance can be determined graphically by a reproduction of Fig. 35, and be transferred with dividers to the projection (Fig. 34). (The beginner will find it easier at first to use a separate auxiliary diagram for such constructions,

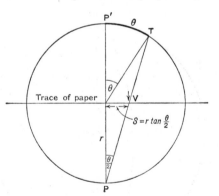

FIG. 35. Derivation of the value of S in stereographic projection.

with a circle of radius equal to that of the primitive, though the practised crystallographer makes use of the primitive itself for this purpose, see construction 2, p. 27.) The four upper faces lie symmetrically disposed around the vertical tetrad, at this determined distance from the centre, and the remaining four beneath are represented by rings around these four dots, since the lower part of the crystal is similar to the upper half.

The twelve faces of the new form have now been projected (Fig. 34); each face is normal to one of the diad axes of the cubic symmetry group in question, and the form is therefore the rhombic dodecahedron (Fig. 11). The next step is the projection of the octahedral faces. From the parallelism of edges, again, it is evident that each octahedral plane lies in a zone with a face of the cube on the one side and a face of the rhombic dodecahedron on the other. (In the investigation of an actual crystal, instead of a model, this zonal relationship would be revealed in the course of the goniometrical measurements.) The octahedral poles can therefore most easily be inserted in the projection by locating on it the intersections of the traces of these zones. Their intersections with the sphere, however, are in four instances great circles inclined to the plane of the projection, and we must first discuss one of the fundamental properties of stereographic projection —that *any circle drawn upon the sphere will be projected as a circle.* (It is interesting to note that this property was clearly realised as early as the thirteenth century.)

We have seen that great circles are of special importance in crystallography, since the poles of all faces in a zone will lie on a great circle.

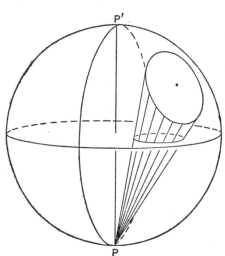

FIG. 36. Illustrating the stereographic projection of a small circle drawn on the sphere.

The property, however, is a general one and we shall prove it by consideration of the general case of a *small circle* (Fig. 36). (A small circle is a section of the surface of the sphere by a plane not passing through the centre of the sphere; parallels of latitude are small circles on the terrestrial globe.) Fig. 37 is a vertical section of the sphere, V is the centre of the small circle on the sphere, and LM the trace of its plane in the section. The right section (through LN) of the cone of which LPN is a cross-section is therefore an ellipse, and LM is the trace of one of its circular sections. Symmetrically inclined to the axis PV in such a cone there is a conjugate circular section QN. Draw MR parallel to the plane of the projection; then

$$\angle RMP = \angle PLM \text{ (on equal arcs } RP, PM)$$
$$= \angle PNQ \text{ (since } LM, QN \text{ are conjugate sections).}$$

Hence MR, and therefore the plane of projection, is parallel to the circular section QN of the

cone. Parallel sections of a cone are similar, and therefore the section of the cone by the plane of projection is a circle (Fig. 36). Incidentally it is clear from the figure that, since the projection of V does not lie halfway between the projections of L and M, the centre of the circle in projection does not coincide with the projection of the centre of the circle on the sphere.

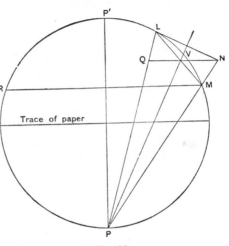

The student familiar with the geometry of inverse loci

FIG. 37.

will observe that this property of stereographic projection follows directly from the proposition that the inverse of a circle is a circle, the locus of the intersections with the projection being the inverse of the circle on the sphere. For, if L be a point on the sphere at an angular distance θ from the summit, and l its stereographic projection, $PL = 2r \cos \theta/2$, and $Pl = r \sec \theta/2$, whence $PL \cdot Pl = 2r^2$.

Returning now to the completion of the projection of the simple cubic crystal, each octahedral pole can be located at the intersection of the traces of three zone-circles (Fig. 38). For each pole, one of these zone-circles in projection is of infinite radius, the others are

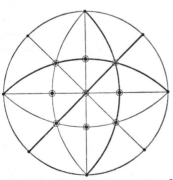

FIG. 38. Completed stereogram of the crystal in Fig. 33. The three zone circles intersecting in one of the octahedral poles are drawn in thicker lines.

circular curves the centre of which in projection can be located by bisecting the chords joining the appropriate poles of the cube and rhombic dodecahedron. (The beginner should note that it is merely an accident of the geometry of a cubic crystal that the centres in projection of these first inclined zone-circles to be drawn happen to lie on the primitive. The centre of a great circle in projection may lie anywhere between the centre of the primitive and a point at infinity, depending on the inclination of the great circle to the plane of projection.) The addition of four rings to represent the four lower octahedral planes completes the *stereogram* of this crystal. In Fig. 39, the arcs of the

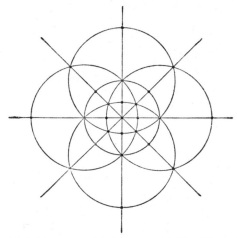

FIG. 39. Completed stereogram of the same crystal (Fig. 33), projected entirely from the lower projecting point *P*.

great circles are continued beyond the primitive to locate these poles in a version of the stereogram in which the lower half of the crystal also is projected from the point *P*; the device of changing to the opposite point *P'*, however, is almost universally used.

STEREOGRAPHIC CONSTRUCTIONS

Since a stereogram will be constantly used as a representation of a given crystal in later discussions, it is opportune to consider here various constructions for which a need will gradually arise. It is not necessary to master them all immediately.

1. Projection of small circles.

(*a*) *About the centre of the primitive.* The stereographic representation of the angular radius is plotted outwards from the centre of the primitive along any diameter, and the circle drawn with its centre in projection at the centre of the primitive. Note that this is the only instance in which the centre of a small circle in projection coincides with the projection of its centre on the sphere.

(*b*) *About a pole within the primitive.* Draw the diameter of the primitive passing through the centre *V* of the required small circle, and on it plot points *L* and *M* at stereographic distances from the given pole on either side corresponding to the angular radius of the required small circle (Fig. 40). *L* and *M* must thus represent the opposite ends

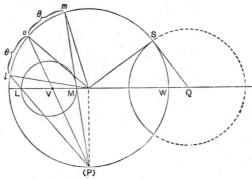

FIG. 40. The stereographic projection of small circles.

of a diameter of the small circle, and the centre in projection must lie at the mid-point of *LM*. If the value of the stereographic distance from the centre for the point *L* exceeds *r*, the radius of the primitive *L* must of course be plotted beyond the primitive (i.e. in Fig. 35 the projecting point *P* is used throughout, whether the point on the sphere lies above or below the plane of the paper).

(*c*) *About a pole on the primitive.* The construction (*b*), extended outside the primitive, covers this case also, but there is a more convenient construction which the student should verify for himself. From the pole *W* on the primitive locate a point *S* at an angular distance along the primitive equal to the radius of the required small circle (Fig. 40). At *S* draw a tangent *SQ* to the primitive, cutting the diameter through the given pole at *Q*. Then *Q* is the centre and *QS* the radius of the required small circle in projection.

2. To find the ' opposite' of a pole. The *opposite* of a pole *V* (Fig. 41) is the projection of the other end of the diameter of the sphere passing through *V*. This is usually projected as a small ring *V'* equidistant from the centre on the side remote from *V*. As the ring indicates, this position arises by changing from the lower to the upper projecting point, and for certain constructions it is necessary to find the true position V'' of the opposite if the lower end of the diameter also is

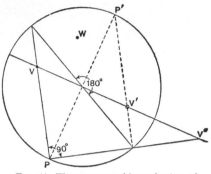

FIG. 41. The stereographic projection of an opposite.

projected from the same projecting point as the upper end.

Suppose that the projecting sphere is rotated through 90° about the diameter of the primitive passing through V; the projecting points P and P' then temporarily occupy positions in the plane of the paper (Fig. 41), the position of V, which is on the axis of rotation, being unchanged. From P project V back on to the trace of the sphere, draw the diameter of the sphere, and project its lower end from P to give the point V''. Restoring the sphere to its original attitude, the position of V'', likewise on the axis of rotation, is unchanged and is the required projection of the opposite of V. Now from the figure it is clear that, since the diameter subtends an angle of 180°, the $\angle VPV''$ at the circumference $= 90°$. Hence V'' can be quickly located in practice by placing a set-square with its right-angle at P and one side passing through V.

3. To draw a great circle through two poles within the primitive. The required great circle passes also through the opposites of the given poles V and W (Fig. 41). One of these opposites is constructed as in construction 2, and a circle constructed passing through the three poles V, W and V''. This is the required great circle, and the arc included within the primitive (the only portion usually represented in projection) will have as chord a diameter of the primitive.

4. To find the pole of a great circle. The *pole* of a great circle is a point on the sphere 90° from every point on the circle; that is, it is the point where the normal to the circle intersects the surface of the sphere. If the great circle be a zone-circle, then its pole is the point of emergence of the zone-axis, regarded as drawn through the centre of the sphere.

The procedure resembles that which we used in construction 2. If ACB (Fig. 42) is the projection of the given great circle, draw the diameter CD of the primitive normal to the chord AB. Project the pole C from B, measure an arc of 90° around the primitive over the pole *A*, and re-project from *B* to give the required pole *D*. From the figure it is clear that, if the angle at the centre = 90° then ∠ *CBD* at the circumference = 45°, and use of the appropriate set-square is again the quickest means of locating the pole *D*.

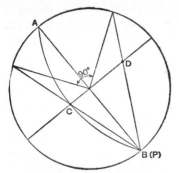

FIG. 42. The stereographic projection of the pole of a great circle.

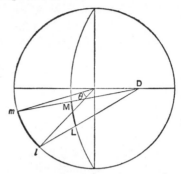

FIG. 43. Measurement of an arc of an inclined great circle.

5. To measure a given arc on an inclined great circle. *L, M* (Fig. 43) are two poles on an inclined zone-circle, and it is required to determine the value of the arc *LM*. Locate the pole *D* of the great circle by construction 4, and from *D* project the poles *L, M* on to the primitive at *l, m*. Then the arc *lm* (measured, of course, by the angle which it subtends at the centre of the primitive) gives the required value.

This construction can also be used for the inverse problem of finding a pole *M* on an inclined great circle at a given distance from a fixed pole *L*. The proof of this property of the pole of a great circle becomes clear when we realise that the straight lines *DLl* and *DMm* are the traces in projection of small circles passing through *D* and the lower projecting point on the sphere. In Fig. 44 the great circle *Al'n'*

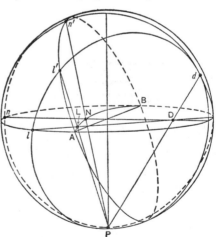

FIG. 44. Spherical projection to explain the construction shown in Fig. 43.

intersects the great circle *Aln* (the primitive of the stereographic projection) in the diameter *AB*. If *d* is the pole of the circle *Al'n'* and *P* the pole of the circle *Aln*, then any circle, great or small, such as *dl'lP* passing through *d* and *P* cuts off equal arcs *Al'*, *Al* on these two great circles. Hence *Al'* = *Al*, *An'* = *An*, and therefore *l'n'* = *ln*. In stereographic projection, the arc *l'n'* projects as the arc *LN*, the pole *d* projects to *D* and the circles *dl'l*, *dn'n* project as the straight lines *DLl*, *DNn*. Hence *ln* is a measure of the arc *LN*; in general, the arc *ml* projected on the primitive (Fig. 43) is a measure of the arc *ML* on an inclined great circle.

GNOMONIC PROJECTION

Neumann described the adaptation to crystallographic work of other methods of projection in addition to the stereographic. Of these, the most important was the *gnomonic* projection, first used in astronomy in the seventeenth century. Little use was made of this method of projection, however, until Mallard * employed it extensively two hundred years later. Its relationship to stereographic projection is illustrated in Fig. 45. The plane of projection, instead of passing through the centre

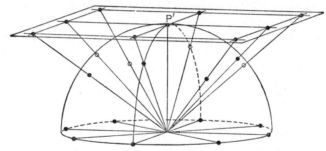

FIG. 45. The principle of gnomonic projection, and the relationship of a gnomonogram to a stereogram.

of the sphere, is tangential to the sphere at the uppermost point *P'*. Each face is projected by direct extension of its normal to intersect the plane of projection. The distance *G* of any gnomonic pole from the centre of the projection is thus $r \tan \theta$, where *r* is the normal distance of the origin from the plane of projection (the radius of the sphere in Fig. 45). Fig. 46 is a *gnomonogram* of the cubic crystal represented in Fig. 33.

One special advantage of this method of projection is at once

* François Ernest Mallard was born at Chateauneuf, central France, in 1833. He became a mining engineer, and was later appointed Professor of Mineralogy at the School of Mines in Paris. He published a large number of papers on crystallography and mineralogy; his best-known work is a *Traité de Cristallographie*, of which three volumes were planned but only two appeared (1879, 1884). He died in Paris in 1894.

apparent. The normals to a set of faces in a zone lie in one plane, and therefore when produced to intersect the plane of projection will give a series of collinear poles—the zone-circles of the stereogram are represented by straight zone-lines in the gnomono-gram. On the other hand, small circles will project as conic sections, which are not easily reproduced graphically; poles of faces normal to the plane of projection lie at an infinite distance from the centre of the projection; and the projection is not angle-true (angles between zone-lines on the projection are not equal to the angles between the corresponding zone-axes). For these reasons, the gnomonic

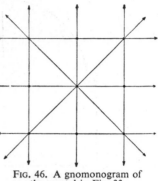

FIG. 46. A gnomonogram of the crystal in Fig. 33.

method is not so generally useful in elementary work as is the stereographic, and we shall make little direct use of it at present.

CYCLOGRAPHIC AND LINEAR METHODS

Two further methods of representation were suggested by Neumann. In the *cyclographic* method, each face is represented by the stereographic trace of a parallel plane passing through the origin (in Fig. 42, a face-pole at D would be replaced by the trace of the great circle ACB). Since the face is thus represented by a plane instead of a point, this method is specially applicable to certain problems in crystal optics and other branches of crystal physics, in which it is necessary to reproduce in projection lines drawn on the crystal face.

A *linear* projection is related to the usual type of gnomonogram in the same way as the cyclographic is related to the usual type of stereogram. If the origin-planes, parallel to the crystal faces, are extended to intersect a non-central plane of projection, each face can be represented by a face-line thus obtained. Little use is made of this projection at the present time.

AUXILIARY DEVICES FOR GRAPHICAL WORK

If much graphical work is to be carried out, a number of helpful devices soon suggest themselves. Thus if we restrict our projections to a constant radius ($2\frac{1}{2}$ ins. will be found generally convenient, or 10 cms. for more accurate work) the edge of a square protractor can be graduated to correspond to distances $r \tan \theta/2$ from $\theta = 0°$ to $\theta = 90°$

(Fig. 47) for stereographic work, or to *r* tan θ for use in gnomonic projections. Among a number of devices of this kind which have been

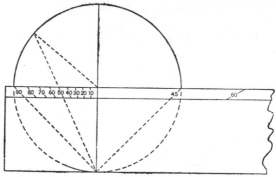

FIG. 47. The principle of the stereographic protractor.

proposed from time to time, the *Hutchinson * protractor* is perhaps the best combination of utility with simplicity. It combines a stereographic and a gnomonic scale, and a special method of graduation affords useful help in a number of constructions.†

A *stereographic net* on which great and small circles of various radii are drawn (Fig. 48) is also useful. Printed on tracing paper, it can be

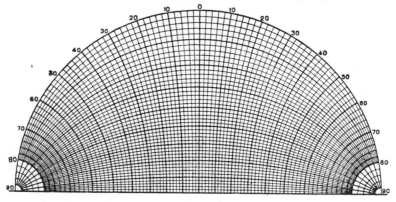

FIG. 48. A stereographic net.

* Arthur Hutchinson was born in London in 1866. After graduating at Cambridge he studied at Würzburg and Munich. Returning to Cambridge in 1891 to a teaching post in Chemistry, he became in succession a demonstrator, a lecturer and (1926) Professor of Mineralogy. A highly successful teacher, he played a prominent part in the development of the Department of Mineralogy. Many of his publications were concerned with graphical and instrumental methods in crystallography. He died in Cambridge in 1937.

† Obtainable from W. H. Harling, 117 Moorgate, London, E.C. 2.

placed over the projection and used both for measurement and for plotting (the required pole being pricked through on the stereogram). In more advanced work, this arrangement may conveniently be reversed, and the projection be carried out on tracing cloth rotated above a printed net. Such a stereographic net is often called a Wulff net, after G. V. Wulff,* who published a reproduction of a net 20 cm. in diameter in 1902; similar devices had been used much earlier by other crystallographers, and an excellent reproduction of a net accompanies a work on astronomy published early in the seventeenth century.

* G. V. Wulff was a Russian crystallographer, whose name should strictly be transcribed as Yurii (Georgii) Viktorovich Vulf. Born at Nezhin in 1863, he graduated at Warsaw University, where he later held the Chair of Crystallography and Mineralogy. In 1907 he was appointed Professor of Crystallography at Moscow University. Many of his papers on crystallography and crystal optics, written originally in Russian, were translated into other languages and in the German translations he used the version of his name, G. Wulff, by which he is usually known outside Russia. He died in Moscow in 1925.

CHAPTER III

THE DESCRIPTION OF CRYSTALS

THE WORK OF HAÜY AND THEORIES OF CRYSTAL STRUCTURE

So far we have thought scarcely at all about possible reasons for the external regularities exhibited by the crystals which we have been measuring, but even the earliest crystallographers naturally began to speculate about the probable internal structural arrangements. Guglielmini made observations on the constancy of *cleavage* directions in a given substance—planes in the crystal along which it could be caused to split regularly—and he believed that the ultimate units from which a crystal is built up must themselves be miniature crystals with plane faces. Nearly a hundred years later, in 1784, the Abbé Haüy * published a work entitled *Essai d'une théorie sur la structure des crystaux appliquée à plusieurs genres de substances crystallisées*, setting forth ideas to which he, also, had been led by observations on cleavage. By an accident during the examination of a group of crystals of calcite (trigonal $CaCO_3$) in the mineralogical collection of an amateur and friend, one of the larger crystals broke off. It seems clear that Haüy himself (by a ' fortunate awkwardness ' in view of all that sprang from the accident) had let the group fall, but his forgiving friend presented him with the broken crystal. ' The prism had a single fracture along one of the edges of the base, by which it had been attached to the rest of the group. Instead of placing it in the collection which I was then making, I tried to divide it in other directions, and I succeeded after several trials in extracting its rhomboid nucleus.'

Finding thus that a rhombohedral cleavage nucleus could be extracted from this particular crystal, he was led to experiment in turn with calcite crystals of other habits, and found that the shape of the rhombohedral unit obtainable by cleavage was constant, and independent of the external habit of the crystal from which it was obtained. After many similar experiments on other substances, he propounded

* René Just Haüy was born in St. Just, northern France, in 1743, and studied in Paris at the College of Navarre and the Cardinal Lemoine College. From the study of physics he turned to Mineralogy, and published a four-volume *Traité de minéralogie* in 1801, whilst keeper of the cabinet at the School of Mines. He was later appointed Professor of Mineralogy in the Museum of Natural History, and ultimately filled a Chair in the new University of Paris. As an honorary canon of Notre Dame he is usually known as the Abbé Hauy. He died in Paris in 1822.

the view that continued cleavage would ultimately lead to a smallest possible unit, a *molécule intégrante*, by a repetition of which the whole crystal is built up. External faces parallel to the directions of cleavage are easily reproduced by regular stacking of the units, but Haüy went on to show how faces with other slopes could be supposed to be formed by omitting rows of units regularly in successive layers. Fig. 49 is a reproduction of one of his figures illustrating the construction of the 'dog-tooth' habit of calcite from rhombohedral units.

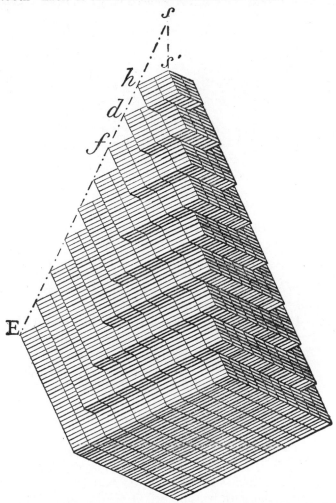

FIG. 49. A reproduction of one of Haüy's figures, showing how a crystal of dog-tooth spar may be considered to be built up from rhombohedral units.

The shape of the fundamental unit was supposed by Haüy to be one appropriate to the particular system of symmetry to which the substance belonged, and we may imitate his experiments in the cubic system by stacking small unit cubelets. A crystal of simple cubic habit corresponds to straightforward orthogonal stacking of cubelets, but by omitting rows regularly as we proceed it is possible to develop planes parallel to the faces of the rhombic dodecahedron. This method of stacking is illustrated in Fig. 50, whilst Fig. 51 shows the completed

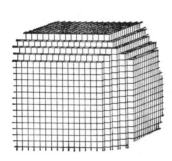

FIG. 50. Stacking cubelets to develop faces of the rhombic dodecahedron.

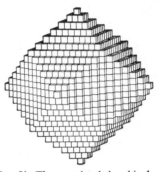

FIG. 51. The completed rhombic dodecahedron built from cubelets.

rhombic dodecahedron. In a similar manner, by regular omission of cubelets from the corners of the stack, the octahedron may be reproduced (Fig. 52). The faces on these forms are, of course, ' stepped ',

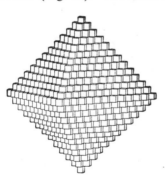

FIG. 52. An octahedron built by stacking cubelets.

but this appearance of roughness is due to the relatively large size of the unit cubelets used, and if we suppose with Haüy that the units in actual crystals are of submicroscopic dimensions the new faces would be apparently plane.

There are many difficulties, however, in the way of accepting Haüy's ideas in the form in which he originally put them forward. In many crystals it is not possible to obtain a structural unit by cleavage; a substance may show no cleavage at all, or only a single direction of cleavage. In a cubic substance cleaving parallel to the faces of the octahedron the unit must be supposed to be octahedral or tetrahedral, and a series of such units cannot be stacked to fill space, though Haüy developed other means of attack by which to discover the probable structural unit in such cases. More seriously, as the atomic theory of matter developed it appeared likely that the interior of a crystal is only partly occupied by solid matter, that much of it is ' empty space '.

Wollaston, in 1812, suggested that the unit cubelets in cubic crystals should be replaced by inscribed spheres, and in this suggestion was reviving views put forward by the versatile Robert Hooke * as early as 1665, when he wrote in his *Micrographia*: ' There was not any regular Figure, which I have hitherto met withall, of [Metals, Minerals, Precious Stones, Salts and Earths] that I could not with the composition of bullets or globules, and one or two other bodies, imitate, even almost by shaking them together.' Even this picture does not help greatly in the case of substances in which we must suppose there are atoms of several kinds, or groups of atoms united to form molecules (which can scarcely be pictured as spheres). Yet such difficulties did not prevent Haüy from making the most important single discovery in crystallography, which justly entitles him to be considered ' the father of crystallography '. We are only concerned at present with the geometrical consequences of the existence of a fundamental unit, and we can avoid these difficulties by replacing each unit by a representative point such as its centre of gravity. If we join adjacent points (Fig. 53) a new unit is outlined, but it is now a structural unit—a unit of pattern—and we make no specific statement about its contents except that it must be supposed to contain a complete representation of the substance of the whole crystal.

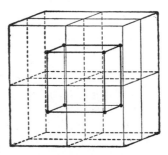

In our discussion of the simplest forms of a cubic crystal it was observed that a face of the rhombic dodecahedron occurs in a zone between two cube faces (Fig. 54), and makes an angle of 45° with each of these. Suppose that a further new

FIG. 53. Development of a structural unit to replace solid cubelets.

face is developed in this zone; will it likewise cut the cube-dodecahedron edge symmetrically, so that it makes angles of $22\frac{1}{2}°$ with each of these faces? Haüy's work suggests that the answer is negative; the new face

* Robert Hooke was born in the Isle of Wight in 1635. From 1662 to the end of his life he was curator of experiments to the Royal Society, and he also held a professorship of geometry in Gresham College. His alert and enquiring mind led him to investigate a wide range of physical phenomena, and the pages of his *Micrographia* make fascinating reading. His name is most familiar in application to the Law *Ut tensio sic vis*. He died in London in 1703.

will slope at an angle determined by the type of stacking which develops it, missing two rows in one direction for one in the other (Fig. 55). It

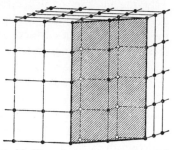

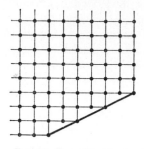

FIG. 54. The relationship of a face of the rhombic dodecahedron (shaded plane) to the structural units.

FIG. 55. Plan of the development of a further face modifying a cube edge.

is therefore the *tangent of the angle of slope* which is halved, and the angle the tangent of which is $\frac{1}{2}$ has the value $26° 34'$ (Fig. 55). Measurement of a crystal, (of calcium fluoride CaF_2 for example), showing this form will confirm this predicted value.

If the crystal in question be one of lower symmetry the underlying unit must also be appropriately less symmetrical. An orthorhombic crystal, for example, will be supposed to be built with rectangular parallelepipeda (the shape of the ordinary brick), but the appearance of new forms is still determined by the regular steps of the stacking. The particular dimensions of the unit, in such a system of lower symmetry, depend on the particular substance, and we cannot predict beforehand the value of a particular angle. From one measured angle, however, we can predict the value of others, for the tangents of the angles of slope are still simply related (Fig. 56). Moreover, as Haüy himself pointed out, the total number of different slopes found in actual crystal faces tends to be small, corresponding to simple variations of the method of stacking. Still treating the problem in two dimensions for simplicity, the full lines of Fig. 56 represent slopes corresponding to the three schemes:

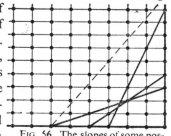

FIG. 56. The slopes of some possible planes in a crystal based on an orthorhombic structural unit.

Miss two units to the right, and miss one upwards.

  ,, one ,, ,, ,, ,, one ,,

  ,, one ,, ,, ,, ,, three ,,

The dotted line corresponds to the more complicated scheme of missing three to the right and five upwards, and it will be seen that the significant points are very sparsely distributed in such a plane, suggesting that it may have little importance as a possible face.

We can illustrate these points by a study of the crystal of orthorhombic barium sulphate, $BaSO_4$, shown in Fig. 57. In the prominent

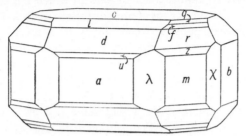

FIG. 57. A crystal of barium sulphate.

zone $a\lambda m\chi b$ the edges am and mb are modified by smaller faces of the forms λ and χ; measuring the angles which these faces make with a, and taking the tangent of the angle am as a standard for comparison, the following results are obtained:

Interfacial angle	Tangent	Ratio
$a\lambda = 22° \ 10'$	0·407	$\frac{1}{2}$
$am = 39° \ 11'$	0·815	1
$a\chi = 67° \ 45'$	2·444	3
$ab = 90° \ 0'$	∞	∞

The particular value of the angle am is evidently determined by the shape of the structural unit of barium sulphate, but the faces λ and χ in the same zone slope at angles the tangents of which are simply related to the tangent of this angle am. Similar results are obtained from other zones:

Interfacial angle	Tangent	Ratio
$au = 31° \ 49'$	0·620	$\frac{1}{2}$
$ad = 51° \ 9'$	1·241	1
$al = 68° \ 3'$	2·481	2
$ac = 90° \ 0'$	∞	∞
$mz = 25° \ 41'$	0·481	$\frac{1}{2}$
$mr = 43° \ 54'$	0·962	1
$mf = 55° \ 17'$	1·443	$\frac{3}{2}$
$mq = 62° \ 32'$	1·924	2
$mc = 90° \ 0'$	∞	∞

THE LAW OF RATIONAL INDICES

These considerations lead us, as they led Haüy, to the formulation of the *Law of Rational Indices*. This, the most important law of crystallography, is perhaps also the most difficult for the beginner to grasp correctly. We shall consider its formulation in three stages.

First, we have to choose a set of axes to which the geometry of the crystal can be related after the manner of use of axes in solid geometry. The simple relationships between slopes of faces which have been deduced above obviously depend on measuring the intercepts made by various planes *in terms of* the lengths of the edges of the underlying structural unit, though certain alternative choices of direction would have given an equally simple result. In Fig. 55, for example, the diagonals of the square units would serve just as well as the sides.

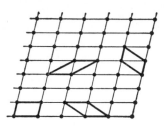

FIG. 58. Various choices of a unit parallelogram in a given pattern.

Whilst therefore the set of axes chosen must be *related to* the internal structural unit, it will not always be possible to choose them unambiguously from the external appearance of the crystal. Often, indeed, as in a crystal founded on a pattern of parallelograms (Fig. 58), there is no unique structural unit. The parallelograms outlined are all equal in area, and any one would serve to reproduce the given pattern of dots. These points can be summarised in the formal statement:

I. **Choice of reference axes.** Any three straight lines, not in the same plane, parallel to actual or possible edges of the crystal. (It is convenient where possible to choose these parallel to prominent axes of symmetry.) They are called *crystallographic axes*; note that they cannot necessarily be chosen orthogonal.

We next proceed to choose a unit plane to define the units of measurement to be employed when measuring along each of the crystallographic axes. This plane is called the *parametral plane*, since it defines the units or *parameters* for the crystal. Its choice presents little difficulty; a plane parallel to any observed face on the crystal will serve, provided that it cuts all three axes (when extended, if necessary):

II. **Choice of parametral plane.** Any plane, parallel to a crystal face, which is not parallel to any of the crystallographic axes. Let it make intercepts *a*, *b* and *c* on these axes.

We are now in a position to describe the slopes of any other observed faces in terms of the unit intercepts thus determined:

III. Definition of Indices. The intercepts made on these axes by any other face *can be expressed as* $\frac{a}{h}, \frac{b}{k}, \frac{c}{l}$ where h, k, l are simple rational numbers or zero. These numbers are termed the *indices* of the face.

The Law of Rational Indices is implicit in statement III, that the indices of any face, thus defined, are always rational. This particular crystallographic notation (essentially the only one surviving in present usage) was popularised by the work of W. H. Miller (p. 20), though the first conception of it appears in works by earlier authors, and hence the numbers are known as the *Millerian Symbol* of the face. The three figures are written without intervening commas (except in the rare cases involving double figures), but are always read separately. Thus 411 is read as four one one; 100 as one nought nought. The indices of a particular face are written, and printed, either unenclosed—as 321 —or enclosed in smooth brackets—(321). (In X-ray crystallography, concerned with ' reflection ' by structural planes in various orientations, it has become a convention to use the unbracketed symbol to denote a family of structural planes and to bracket the index to denote the actual crystal face. Thus 110 ' reflections ' arise from structural planes parallel to the face (110)). If the symbol is enclosed in braces —{321}—it denotes all the faces of the *form* generated from the face (321) by the operation of the symmetry in question. The special significance of square brackets or crotchets—as [321]—will be described later (p. 203).

Considering the formulation set out above, it is clear that the following is the rule for determining the indices of a particular face:

> Divide the intercepts made by the face into the standard parametral intercepts *a, b, c*. Multiply the result, if necessary, to clear of fractions.

Applying this result to the parametral plane itself, the indices will be a/a, b/b, c/c, i.e. 111; this is the symbol of the parametral plane in any crystal, whether the symmetry be cubic or triclinic, whether the intercepts it makes on the axes be all equal or unequal. The index 111 merely shows which plane was chosen to determine the units, and since different observers may make different choices it is possible to have different descriptions of a single crystal, all equally correct.

In Fig. 59, *ABC*, *DBC* and *DBE* represent the slopes of three faces in a crystal the structural unit of which, with edges of lengths

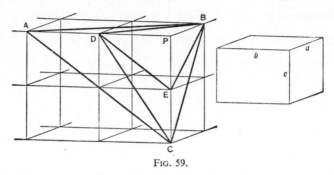

FIG. 59.

a, *b*, *c*, is represented on the right of the figure. Suppose that, for this particular substance, the ratios $a : b : c$ of the actual unit $=0.816 : 1 : 0.924$. (It is customary to express the ratio $a : b : c$ in the form $a/b : 1 : c/b$, reducing *b* to unity.)

The first observer selects the plane *ABC* as parametral plane; he therefore assigns to it the index 111. From its slope in relation to the three axial directions he would determine *axial ratios* of values $BP : AP : CP$, $0.408 : 1 : 0.924$.

The index of the plane *DBC* is $\dfrac{BP}{BP} \dfrac{AP}{DP} \dfrac{CP}{CP}$, i.e. 121.

„   „   „   *DBE* is $\dfrac{BP}{BP} \dfrac{AP}{DP} \dfrac{CP}{EP}$, i.e. 122.

The second observer selects the plane *DBC* as 111, and hence calculates axial ratios $BP : DP : CP$, $0.816 : 1 : 1.848$.

The index of the plane *ABC* is $\dfrac{BP}{BP} \dfrac{DP}{AP} \dfrac{CP}{CP}$, $1\frac{1}{2}1$, i.e. 212.

„   „   „   *DBE* is $\dfrac{BP}{BP} \dfrac{DP}{DP} \dfrac{CP}{EP}$, i.e. 112.

To the third observer, *DBE* seems the best choice as 111. His calculated axial ratios, $BP : DP : EP = 0.816 : 1 : 0.924$, are actually those of the edges $a : b : c$ of the structural unit.

The index of the plane *ABC* is $\dfrac{BP}{BP} \dfrac{DP}{AP} \dfrac{EP}{CP}$, $1\frac{1}{2}\frac{1}{2}$, i.e. 211.

„   „   „   *DBC* is $\dfrac{BP}{BP} \dfrac{DP}{DP} \dfrac{EP}{CP} \dfrac{}{1\frac{1}{2}}$, 1, 221. i.e.

Tabulating these results, we have three correct but different descriptions of the slopes of the three planes present on the crystal:

	Axial ratios	Indices of planes		
		ABC	*DBC*	*DBE*
Observer 1 -	0·408 : 1 : 0·924	111	121	122
Observer 2 -	0·816 : 1 : 1·848	212	111	112
Observer 3 -	0·816 : 1 : 0·924	211	221	111

Each description illustrates the rationality (and the simplicity) of the indices, but only when we have advanced so far that we can determine the exact arrangement of the internal structure, and hence the absolute lengths of the edges *a*, *b*, *c* of the unit of structure, could we say that the description given by observer 3 is preferable to those offered by observers 1 and 2.

It is clear, also, that the rationality lies only in the *ratios of intercepts* on corresponding axes, and there is no rational relationship between the intercepts on different axes unless by chance the symmetry of the system permits the choice of a symmetrically-situated plane as parametral plane. Hence the name *Law of Rational Intercepts* sometimes used is particularly misleading; it can be corrected by writing *Law of Rational Ratios of Intercepts*, but as a ratio of intercepts defines an index the correct name of the Law is that which we have used above.

Finally, it must be observed that there is nothing in this discussion which determines the relative dimensions of the whole crystal; we have been concerned only with the slopes of faces, and the habit of the crystal is in no way necessarily related to the units of measurement *a b c* which determine these slopes. All these points will be much clearer when we have been able to practise making accurate drawings of crystals.

CRYSTAL DRAWINGS

The three crystallographic axes are denoted by the letters *x*, *y*, *z*, and in crystal drawings the positive direction of the *x* axis is always regarded as running forwards more or less towards the observer, positive *y* towards the right, and positive *z* upwards parallel to the margins of the paper. An intercept made on the negative portion of any axis is denoted by a minus sign placed over the corresponding figure in the index. Thus the plane drawn in Fig. 60 is $h \bar{k} l$ (aitch minus-kay ell). We shall begin the consideration of the construction of crystal drawings in terms of the cubic system, and in this system it is possible so to choose the parametral plane that the axial units are all equal, $a : a : a$.   To construct an *axial cross* for the cubic system we

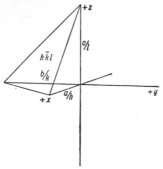

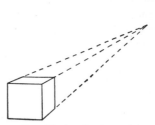

FIG. 60. The conventional orientation of crystallographic axes.

FIG. 61. A cube drawn in
perspective.

therefore require a representation of three equal straight lines at right angles.

When an artist draws a picture of a cube, he represents it in some such way as Fig. 61; drawing ' in perspective ' he makes edges which run away from the observer converge towards a point, although he is well aware that these edges are actually parallel on the cube itself. Since parallelism of edges on a crystal reveals the important zonal relationships, we shall clearly not adopt any such convention as this. Whatever the particular method of projection chosen, it must be one of parallel perspective so that all edges which are parallel on the actual crystal are drawn as parallels on the paper.

If a cube be held with its y and z axes parallel to the plane of the paper its elevation projects by parallel perspective on to the paper as a square.   Such elevations (and plans projected parallel to the z axis) were widely used by the early crystallographers; they are, for instance, almost the only kind of illustration in Brooke and Miller's *Mineralogy*. To most readers, however, a ' three-dimensional ' representation is more satisfactory than separate plan and elevations; Haüy used this type almost always, and there is little doubt that the excellence of his illustrations was a great help in gaining rapid acceptance of his views. If the cube, held as above, is now rotated through a small angle θ about the z axis towards the left, the right-hand vertical face becomes visible (Fig. 62); the x axis, foreshortened to a point in the elevation, is represented as a length depending on the value θ, but it is still collinear in projection with the y axis (which has been slightly foreshortened in projection by the rotation). Finally, to reach a 'three-dimensional' figure we can either tilt the cube towards us through an angle ϕ (the lines of projection remaining normal to the paper) to give an *orthographic projection*, or raise the point of view so that the lines of projection are inclined at the required angle ϕ to the normal to the paper, giving a *clinographic projection*.

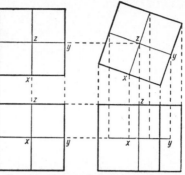

Early drawings on an axial cross were made by orthographic projection, but this method is now little used (though we shall later employ orthographic projection in a drawing procedure which dispenses with the axial cross). The second method was popularised by Nau-

FIG. 62. Plans and elevations of a cube viewed by parallel projection. In the left-hand elevation the direction of view is parallel to the x-axis, whilst the right-hand elevation shows the effect of a small angular rotation about the z-axis.

mann,* who published a simple construction for a clinographic projection of an axial cross. The appearance to the casual observer of drawings

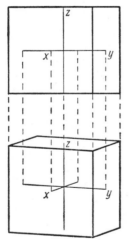

FIG. 63. Development of an orthographic projection of the cube.

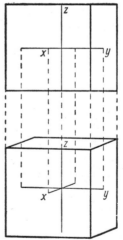

FIG. 64. Development of a clinographic projection of the cube.

* Carl Friedrich Naumann was born in Dresden in 1797, and studied at Freiberg, Leipzig and Jena. In 1826 he became Professor of Crystallography at Freiberg, but soon extended his interests over a wider geological field, and in 1842 moved to the Chair of Mineralogy and Geognosy at Leipzig. He published a number of text-books on crystallography and mineralogy. His death took place in Dresden in 1873.

made by these two methods is closely similar, as shown in Figs. 63 and 64, the former representing a cube on orthographic projection, whilst the latter is constructed on clinographic axes.

CONSTRUCTION OF AN AXIAL CROSS IN CLINOGRAPHIC PROJECTION

(a) **General construction.** Let θ be the chosen angle of rotation about the z axis, and let ϕ be the angle of elevation of the line of projection above the normal to the paper.

Draw a vertical line AOB (Fig. 65) equal in length to $2a$, where

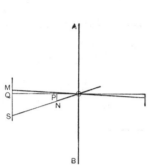

FIG. 65. Construction of an axial cross in clinographic projection.

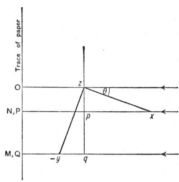

FIG. 66. Plan of the construction of Fig. 65 on a plane normal to the z-axis. After rotation about the z-axis through an angle θ, the negative y-axis would project orthographically as a length OQ on the paper ($=zq=a\cos\theta$) and the x-axis as a length $OP(=zp=a\sin\theta)$.

a is the chosen length of the axes in space. At its mid-point O draw a normal to AB, and mark off $OQ =a\cos\theta$. Along OQ mark off $OP=a\sin\theta$.

At Q draw a normal to QO upwards, and mark off on it

$$QM =a\sin\theta\tan\phi.$$

At P draw a normal to QO downwards, and mark off on it

$$PN=a\cos\theta\tan\phi.$$

Join MO and NO.

Then N, M, A represent unit points on the required x, $-y$ and z axes, at equal distances a from the origin.

(b) **Simplification following Naumann.** The above construction can be simplified by the choice of convenient standard values for θ and ϕ. Naumann proposed the values $\theta=18°\ 26'$ (tan $\theta=\frac{1}{3}$) and $\phi=6°\ 20'$

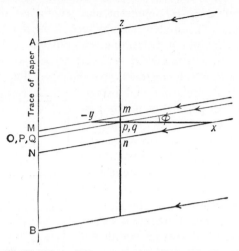

FIG. 67. Elevation of the construction of Fig. 65. Clinographic projection at an angle ϕ above the plane of Fig. 66 projects $-y$ to a point $M(QM=qm=a\sin\theta\tan\phi)$ and $+x$ to a point $N(PN=pn=a\cos\theta\tan\phi)$.

(tan $\phi=\frac{1}{9}$), and these have been extensively employed, though it is customary nowadays to work with a rather larger ϕ value, such as $\phi=9°\,28'$ (tan $\phi=\frac{1}{6}$). Following Naumann, we can propose a simplified construction for the axial cross:

Begin by drawing the base-line OQ. Mark off $OP=\frac{1}{3}OQ$, and from P draw a normal $PN=\frac{1}{3}PO$. Join NO. At Q draw a normal $QM=\frac{1}{3}PN$. Join MO. Draw OA normal to OQ, and mark off $OA=OS$, where $QS=OP=\frac{1}{3}OQ$.

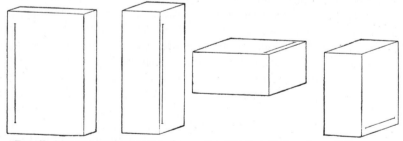

FIG. 68. Two clinographic views of a rectangular parallelepiped $3\times2\times1$. A line is drawn on one of the largest faces parallel to the longest edges. When the parallelepiped is turned to bring the intermediate edges parallel to the x-axis foreshortening makes the cross-section seem almost square.

FIG. 69. Two further views of the same parallelepiped as Fig. 68, with the longest edges set parallel to the x axis.

Then *NO*, *MO* and *AO* represent unit lengths on the required *x*, −*y* and *z* axes. (The student should verify that this particular case agrees with the general construction given above.)

A set of axes thus constructed serves as a basis for the production of crystal drawings which in general give a very satisfactory impression. From the method of construction, however, it is clear that edges parallel or nearly parallel to the *x*-direction are very much foreshortened in the resulting drawing, and the beginner must keep this constantly in mind when attempting to assess the relative dimensions of a crystal in various directions from the inspection of a clinographic drawing. Figs. 68, 69, 70 represent a brick with edges *a*, *a*/2, *a*/3 drawn in six different attitudes in relationship to the *x*, *y*, *z* axes. The line drawn near one edge enables one to compare these attitudes, and it will be seen that if the longest edges are set parallel to the *x* axis (Fig. 69) one receives at first the impression of a very much more nearly cubic object than when these edges lie parallel to the *yz* plane (Figs. 68, 70).

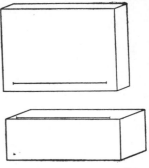

FIG. 70. The same parallelepiped as in the preceding figures, seen with the longest edges set parallel to the *y* axis.

The use of this axial cross in constructing drawings of cubic crystals, and its adaptation to drawings of crystals of lower symmetry, will be developed gradually. We have now discussed in outline the successive steps in the routine crystallographic examination of a substance. We first make a complete set of goniometrical measurements and from these construct an accurate stereogram. From this stereogram the crystallographic symmetry can be determined, an appropriate set of crystallographic axes be chosen, selection made of a parametral plane, and a consistent set of indices be assigned to all the faces present. This involves also the determination of the angles between the crystallographic axes *x y z* and of the axial ratios *a* : *b* : *c*. Finally, an axial cross is constructed, and a drawing made to represent a typical crystal. There is much more to learn in detail about each of these successive steps, but this knowledge is most easily acquired during a more detailed study of each of the crystal systems in turn.

CHAPTER IV

A GENERAL STUDY OF THE SEVEN CRYSTAL SYSTEMS

THE CUBIC SYSTEM, HOLOSYMMETRIC CLASS

Our initial study of this group of symmetry was brought to the stage of inserting on a stereogram (Fig. 38) the poles of the faces of the three forms, the cube, the rhombic dodecahedron and the octahedron. The next step is the choice of the axes $x\,y\,z$. The three directions of cube edge are prominent zone axes, they are parallel to tetrad symmetry axes, and thus provide an orthogonal set of crystallographic axes entirely in accord with the formulation of the Law of Rational Indices. We have seen that in crystal drawings the z axis is set parallel to the margins of the paper; when making the corresponding projection, the z axis is placed to coincide with the line PP' of Fig. 32, with $+z$ upwards, $+y$ horizontal to the right and $+x$ running towards the observer. Hence in the stereogram $+x$ runs downwards parallel to the margins of the paper (Fig. 71), $+y$ towards the right-hand margin, and $+z$ normal to the plane of the paper.

There is in the projection at present only one plane so situated that it can be chosen as parametral plane—an octahedral plane—and to it we assign the indices 111. From the symmetry it is clear that this plane makes equal intercepts on all three axes, so that the *axial ratios* are 1 : 1 : 1 (the parametral units are equal along the three axial directions). Against the pole of the upper right-hand front face of the octahedron (cutting all three axes positively) we write the index 111, and to the other three poles of the same form on the upper hemisphere we allot the indices $1\bar{1}1$, $\bar{1}\bar{1}1$ and $\bar{1}11$ (Fig. 71). It is not usual in a stereogram to write indices against the rings representing faces on the lower hemisphere (unless the lower face only is present); the corre-

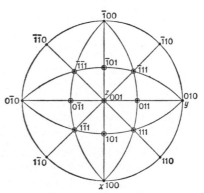

FIG. 71. Indexed stereogram of the crystal in Fig. 33.

sponding index is clearly that of the dot above it with the z index made negative (e.g. the face below $\bar{1}11$ is $\bar{1}1\bar{1}$) and the insertion of all these indices crowds the stereogram unduly. We have thus indexed the *form* {111}; it is redrawn in Fig. 72, with the corresponding index placed on some of the faces.

To draw this form on the axial cross, we require to draw in position a series of edges between faces the indices of which are now known. Consider first the edge between the faces 111 and $1\bar{1}1$. From the indices, each of these faces passes through the point one unit along

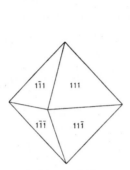

Fig. 72. The octahedron with the front faces indexed.

Fig. 73. Drawing the octahedron on the axial cross.

$+x$, which is therefore a point on the required edge; similarly, each of the faces passes through a point one unit along the $+z$ axis, and the edge is therefore correctly represented by joining unit point on the $+x$ axis to unit point on the $+z$ axis (Fig. 73). Remembering the presence of the triad axes, it is unnecessary to use the indices of other faces for further reasoning; the drawing is completed by repeating this construction symmetrically.

Turning next to the cube, from the positions of the poles of its faces in the stereogram, it is clear that each face is parallel to two of the axes, but intersects the third. Parallelism corresponds to an infinite intercept and the corresponding index is therefore zero; the cube is the form {100} (Fig. 74), and the indices can be inserted in the stereogram (Fig. 71).

Fig. 74. The cube with the front faces indexed.

Each face of the rhombic dodecahedron is parallel to one of the crystallographic axes, but intersects each of the others; the required index is therefore of the type $hk0$. Since the pole of each dodecahedron face lies at 45° between the adjacent cube poles, the intercepts m on these two axes are equal. The indices are therefore $\left\{\dfrac{a}{m}\ \dfrac{a}{m}\ \dfrac{a}{\infty}\right\}$, or in simplest terms $\{110\}$. (Note particularly the exact procedure followed here; the index of a face of the rhombic dodecahedron is not 110 *because* it makes equal intercepts on two of the axes; it is 110 because it makes on these axes intercepts in the same proportion as those made by the parametral plane, and it is an accident of the high symmetry of the system that the parametral units are equal.) Eight of the poles of this form are indexed in Fig. 71; the remaining four poles are on the lower hemisphere.

A drawing of the form is repeated in Fig. 75, with some of the indices inserted on the corresponding faces. To make this drawing we consider as before the indices of faces on either side of a particular edge. The indices 110 and 101 have only one figure in common, giving one point on the required edge at unit distance along the $+x$ axis.

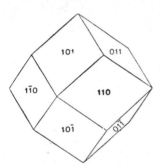

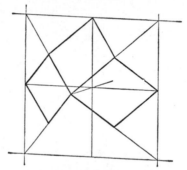

FIG. 75. The rhombic dodecahedron with the front faces indexed.

FIG. 76. Drawing the rhombic dodecahedron.

To obtain another point on the edge, we must reason as follows: from the y and z figures of the index 110 this plane, when extended, would intersect the yz axial plane in a line parallel to the z axis through unit point on the $+y$ axis (Fig. 76), and similarly the crystal plane 101 would intersect the yz axial plane in a line parallel to y through unit point on the $+z$ axis. Drawing these two lines (Fig. 76) they must intersect in a point, in the yz axial plane, which is common to the two crystal planes and therefore lies on the edge between them. Joining this point to the unit point on the $+x$ axis we have constructed the required edge. Since the x axis is a tetrad symmetry axis, the construction may be repeated around it (using in turn unit point on negative y and unit point on negative z) to give the other three directions of edge required. The faces of the rhombic dodecahedron are all rhombuses, which will reproduce in clinographic drawing as parallelograms; the drawing is therefore quickly completed by drawing the appropriate edges parallel to the four constructed lines.

In examining a number of cubic crystals, we should soon encounter specimens of more complex habit than the simple combination of three forms shown in Fig. 33. In what positions are the faces of further forms likely to appear? The predominance of simple zonal relationships which we have already stressed leads us to investigate the possibility of a new form appearing as faces modifying all the edges of the type *ab* in Fig. 33, the edges between adjacent cube and octahedron planes. The most likely shape for such faces is dictated by the probability that its other pair of opposite edges will likewise be parallel, the face falling also in a zone between two adjacent dodecahedral faces. The pole of this face can be inserted in the stereogram (Fig. 77) by

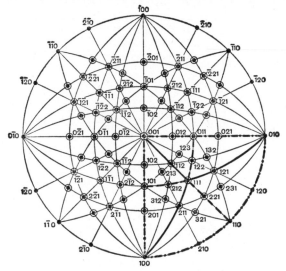

Fig. 77. Stereogram of a cubic crystal.

drawing the zone-circles from 100 to 111 and from 110 to 101; the required pole lies at their intersection. Symmetry demands that there should be three such faces around the triad axis normal to 111, twenty-four faces altogether in the complete form. The symmetrically-modified crystal is illustrated in Fig. 78.

What is the index of this form? The student may have noticed already, in Fig. 71, that there is apparently a simple relationship between the indices of faces falling in one zone. The indices 100 and 010 add up to give the index 110 of a face between them in the same zone; 110 adds up with 001 to give the index 111, and so forth. We shall be able to prove later that this result can be generalised from the law of rational indices; all the poles of possible faces lying in a zone between two given faces *hkl* and *pqr* have indices of the type $mh + np$, $mk + nq$,

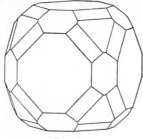

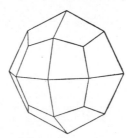

Fig. 78.   Fig. 79. The icositetrahedron {211}

$ml + nr$ obtained by adding m times the index *hkl* to n times the index *pqr*, where m and n are small whole numbers. In the zone between 100 and 111 lies the pole 211, and since this index can also be obtained by adding 110 and 101, it must be the index of the pole at the intersection of these two zones. The new form is {211}; developed alone it has the shape of Fig. 79. We shall call it an *icositetrahedron*; some crystallographers call it a *trapezohedron*, but we shall reserve this name for certain forms without planes of symmetry which will be encountered later in classes of lower symmetry. (A trapezoid is a quadrilateral none of whose sides are parallel, whilst the more familiar term trapezium is applied to any irregular quadrilateral, but especially to one with one pair of opposite sides parallel. It can therefore be maintained that the name trapezohedron is correctly applied to this cubic form, but the restriction which we propose to observe is a convenience in crystallographic nomenclature.)

To draw this form on the axes, we proceed as before from a consideration of the indices of the pair of faces meeting in a particular edge. The figure 2 in the index indicates that some of the intercepts are fractional, but since in clinographic projection there is no foreshortening of distances, the required points are located on the axes, by measurement with a scale, half-way between the origin and the unit points. In Fig. 80 the drawing is being taken a step further, and the simple icositetrahedron is being modified by small faces of the rhombic dodecahedron {110}. To draw such a combination of forms, the icositetrahedron alone is first drawn completely; we then require to find the direction of an edge between an icositetrahedral and a dodecahedral

plane, such as the edge 211–110. Reference to the stereogram (Fig. 77) shows that these planes lie in the dodecahedral zone 110–101, and the direction of the required edge is therefore found by the method of Fig. 76. This being a combination of forms, however, the qualitative problem of the habit arises, and from a point on the icositetrahedral edge chosen to give a dodecahedral face of the desired size an edge is drawn parallel to the constructed direction. The method of completing the drawing will be evident from Fig. 80.

FIG. 80. Drawing the form {211} and modifying it by faces of the form {110}.

Examining the characteristics of the four forms now described, we can see that each owes its shape to the particular relationship of each of its faces to the symmetry elements. Starting with a face normal to a tetrad axis, the symmetry demands six such faces in all, and the cube {100} is thus developed. Similarly, the rhombic dodecahedron {110} arises from the operation of the symmetry on any face normal to a diad axis, and the octahedron from any face normal to a triad axis. Each face of the icositetrahedron {211} is normal to one of the diagonal planes of symmetry, and lies between a cube and an octahedral plane. Yet this particular relationship does not demand that the index need be {211}, for it is true of any face of the type $h l l$ where $h > l$; developed on the crystal of Fig. 33 such a face will have only one pair of opposite edges parallel (Fig. 81), and in projection its poles will lie somewhere

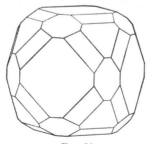

FIG. 81.

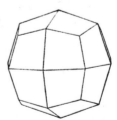

FIG. 82. The icositetrahedron {311}.

on the arcs shown heavily printed in Fig. 77. Thus, whilst there is only one cube, one rhombic dodecahedron, one octahedron, there is a whole family of possible icositetrahedra of which {211} is the simplest. Such forms as {311} Fig. 82, {322} Fig. 83, and {411} are all of the same kind of shape, in this particular crystal class, but differ in the values of corresponding interfacial angles.

A further possibility arises from consideration of faces modifying

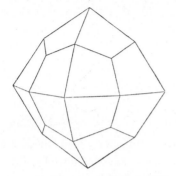 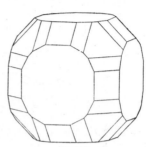

FIG. 83. The icositetrahedron {322}.　　　　FIG. 84.

an octahedron-dodecahedron edge (Fig. 84). Such a face, also, is normal to a diagonal plane of symmetry, but its pole lies between those of octahedron and dodecahedron on one of the zones marked with dashed lines in Fig. 77. The index of the form is of the type $\{h\,h\,l\}$, where $h > l$, and the simplest example is {221}. It is again a twenty-four-faced form, and is conveniently called a *trisoctahedron* (Fig. 85). (This is sometimes written in full as *triakisoctahedron* ; another term,

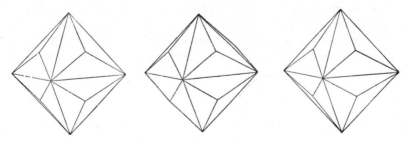

FIGS. 85-87. Trisoctahedra.　{221}　{331}　{332}.

' three-faced octahedron ', should be regarded only as a convenient colloquialism.) Two further members of the family of trisoctahedra are illustrated in Fig. 86—{331}, and Fig. 87—{332}.

The edge between cube and dodecahedron has yet to be modified (Fig. 88).   The index of such a form is of the type $\{h\,k\,0\}$, the simplest example being $\{210\}$.   Each face is normal to one of the cubic planes of symmetry, between the tetrad and diad axes, and the poles lie on

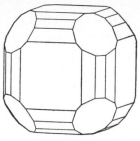

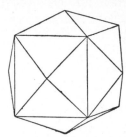

FIG. 88.   FIG. 89. The tetrahexahedron $\{210\}$.

the dot-dash zones of Fig. 77.   Again twenty-four-faced, these forms are *tetrahexahedra* (*tetrakishexahedra*, ' four-faced cubes ').   Fig. 89 illustrates the form $\{210\}$).

The six kinds of form thus far developed—three unique forms arising from faces normal to axes of symmetry, ard three families of forms arising from faces normal to planes of symmetry but not to axes—have used up all the possible positions for a face *specially related* to the elements of symmetry.   Since they show this special relationship these six kinds of form are called *special forms* of this class of symmetry. Any other possible face on the crystal will be quite generally related to the symmetry ; it will not be normal or parallel to an axis of symmetry or to a plane of symmetry, or symmetrically inclined to two axes of symmetry or in fact have any particular kind of attitude in relation to the elements of symmetry which we can formulate in words.   The poles must lie within the spherical triangles outlined in Fig. 77 (the smallest triangle outlined on the sphere by the planes of symmetry is

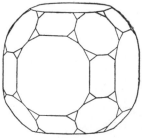

FIG. 90.   FIG. 91. The hexoctahedron $\{321\}$.

sometimes called the *systematic triangle*) and there will be forty-eight faces in all, six in each of the eight octants of space (Fig. 90). The index will be of the type $\{h\,k\,l\}$, and the simplest example is $\{321\}$ (Fig. 91). The appropriate name is *hexoctahedron* (*hexakisoctahedron*, ' six-faced octahedron '), and from its general relationship to the symmetry it is called the *general form* of this class of the cubic system.

Tabulating our results, we have now described the following kinds of form :

Special forms.	Cube	$\{100\}$.	
	Rhombic dodecahedron	$\{110\}$.	
	Octahedron	$\{111\}$.	
	Icositetrahedra	$\{h\,l\,l\}$	$(h>l)$.
	Trisoctahedra	$\{h\,h\,l\}$	$(h>l)$.
	Tetrahexahedra	$\{h\,k\,0\}$.	
General forms.	Hexoctahedra	$\{h\,k\,l\}$.	

There are thus only seven essentially different kinds of possible shapes for individual forms in this class of the cubic system, six of which belong to special forms and one to the general form. We shall see later that the six special forms are not uniquely characteristic of this particular class—the holosymmetric—of the cubic system ; $\{100\}$, for example, is the cube, and $\{110\}$ the rhombic dodecahedron in all the classes of the cubic system ; $\{111\}$ is an octahedron in two further classes, though it is a different shape in some of the remaining classes, and so for the other special forms. The hexoctahedron, however, is uniquely characteristic of this particular class ; only if the symmetry group under consideration comprises the thirteen axes, nine planes and centre of this particular class is the face 321 repeated forty-eight times by the symmetry elements to produce the hexoctahedron as the general form. We can therefore now label this class *the hexoctahedral class* of the cubic system.

Fig. 92 illustrates a crystal of a substance, Fe_3O_4, crystallising in this class, though the specimen illustrated does not show the general form. The predominance of the rhombic dodecahedron determines the habit ; we recognise also small square faces of the cube $\{100\}$ and equilateral triangular faces of the octahedron $\{111\}$. The remaining form shows twenty-four faces, and parallelism of edges reveals that the faces lie in cube-

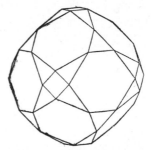

FIG. 92. A crystal of magnetite.

octahedron-dodecahedron zones, between the cube and octahedron faces. The index is therefore of the type $\{h\,1\,1\}$ and the form is an icositetrahedron. It is clearly not the icositetrahedron $\{211\}$, however, since its faces do not lie in a zone with adjacent dodecahedral faces on either side. The particular index could only be determined by measurement.

In Fig. 93, a crystal of spinel, Mg Al_2O_4, of octahedral habit, shows also small faces of the cube $\{100\}$, the rhombic dodecahedron $\{110\}$

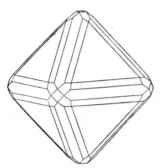

and the icositetrahedron $\{211\}$. The remaining form is shown by its zonal relationships (and hence also by its relationships to the symmetry elements) to be a trisoctahedron $\{h\,h\,l\}$, and the parallelism of edges between the pairs of faces $h\,h\,l$–211 and $h\,\bar{h}\,l$–2$\bar{1}$1 shows that it is the particular trisoctahedron $\{221\}$.

The student should practise drawing on the axes some of the forms discussed above and simple combinations of them. The general procedure to be followed has been

FIG. 93. A crystal of spinel.

outlined in relation to Fig. 80, and we shall add only one further observation here. It may be required to construct the direction of an edge between two faces the indices of which show no figure in common. It then becomes necessary to construct two points each of which is known to lie in the required edge. Suppose (Fig. 94) it is required to find the direction of the edge between the planes 221 and 1$\bar{1}$2. These planes meet the xy axial plane in the lines joining $\frac{1}{2}$ on x to $\frac{1}{2}$ on y and joining 1 on x to $\bar{1}$ on y respectively. Drawing these lines, they intersect to give a point L on the edge. Similarly, a point of intersection M is determined in the xz plane or a point N in the yz plane. The points L, M, N thus obtained are collinear, giving the required direction of edge, and only the two more conveniently obtained in any

FIG. 94. General construction of an edge on the axial cross.

particular case need be located (it is usually best to avoid an intersection in the xy plane, since this plane is so much foreshortened in the standard clinographic projection).

CUBIC CRYSTALS OF LOWER SYMMETRY

Though we shall be chiefly concerned in this chapter with the holosymmetric classes of the various systems, some of the points in the above discussion will be clearer if we examine briefly here some examples of cubic crystals belonging to classes other than the holosymmetric.

Iron disulphide, FeS_2 (the mineral pyrite), frequently crystallises as cubes with characteristically striated faces (Fig. 95). Accepting these striations as an expression of the symmetry of the underlying structure, we find that such a crystal still shows the group of four triad axes which is the mark of the cubic system. The normals to the cube faces, however, are no longer tetrad axes since the cube must be rotated through 180° about any one of these normals before reaching a completely con-

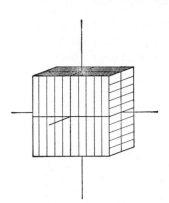

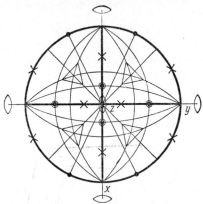

Fig. 95. A striated cube of pyrite.

Fig. 96. Stereogram showing the elements of symmetry of the crystal of pyrite in Fig. 95.

gruent position. There are still three planes of symmetry parallel to the cube faces, but the diagonal planes are no longer planes of symmetry, since they run at an angle to the striations and do not reflect them. The substance belongs to the cubic system but to a lower class of which the symmetry consists of four triads, three diads, three planes and a centre. Marking this symmetry on a stereogram (Fig. 96), we can insert the poles of various specially-situated faces and find how often they are repeated by the symmetry. A face 111, normal to a triad axis, is still repeated eight times, so that the form {111} in this class of the cubic system, also, is the octahedron. Similarly, we should find that {211} and {221} are the same twenty-four-faced forms which we have already described. The face 210, however, is repeated only twelve times, since there is no inclined plane of symmetry or diad axis normal to 110 to produce the face 120 when 210 is given. {210} in this class is therefore a twelve-faced form, the *pentagonal dodecahedron* (Fig. 97). (From its frequent occurrence on crystals of pyrite, this form is sometimes called a pyritohedron ; but since it is a special form its

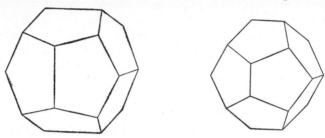

FIGS. 97-98. Pentagonal dodecahedra   {210}   {320}.

occurrence is not confined to crystals of this class and the use of this name is undesirable.) As in the case of the tetrahexahedra of the holosymmetric class, a family of such pentagonal dodecahedra {h k 0} exists, and two further examples are illustrated in Fig. 98—{320}, and Fig. 99—{410}.

If a face occurred in a position 120 on such a crystal it would likewise be repeated twelve times as indicated by crosses in Fig. 96. The form

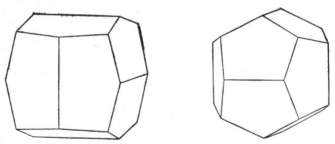

FIGS. 99-100. Pentagonal dodecahedra   {410}   {120}.

{120} is another pentagonal dodecahedron of the same shape as {210}, but with a different attitude in space (Fig. 100). Since each of these forms possesses half the number of faces shown by {210} in the holo-symmetric class, they were termed by early crystallographers *hemi-hedral* forms; the number of faces in a form, however, depends directly on the class of symmetry in question, and here each of the forms {210} and {120} is an independent entity, with the faces of the one showing a different relationship to the underlying structure from those of the other. The term hemihedral (and the related term *tetartohedral*, where a form shows one-quarter of the number of faces in a related holo-symmetric form) should therefore not be used. The general form of this class is a twenty-four-faced form which we shall examine later.

Crystals of cuprous chloride, CuCl, often have the shape of a regular tetrahedron, Fig. 101. Normal to each equilateral triangular face is a triad axis, which emerges through a coign on the opposite side of

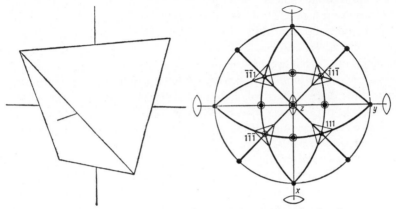

FIG. 101. The tetrahedron {111}.   FIG. 102. Stereogram showing the symmetry of the tetrahedron.

the crystal. An axis of symmetry of this kind is called *uniterminal*, and we are evidently dealing here with an example of crystallisation in yet another class of the cubic system. The symmetry displayed by the tetrahedron consists of four triad axes, three diad axes, and six diagonal

planes (as marked on the stereogram, Fig. 102). The uniterminal character of the triad axes indicates that in this class there is no centre of symmetry. Inserting the pole 111 of the face normal to a triad axis, it is clear that such a pole is repeated only four times ; the form {111} comprises only the four planes 111, $\bar{1}\bar{1}1$, $1\bar{1}\bar{1}$ and $\bar{1}1\bar{1}$, the tetrahedron of Fig. 101. If we had started with the face $1\bar{1}1$ this also would have been repeated four times, giving another tetrahedron {$1\bar{1}1$} (Fig. 103), identical in shape with {111}

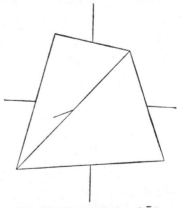

FIG. 103. The tetrahedron {$1\bar{1}1$}.

but differently situated with regard to the axes. These two tetrahedra would formerly have been described as hemihedral forms of the octahedron, but they are to be regarded rather as two special forms appropriate to the class of the cubic system under consideration. Some authors refer to {111} as a *positive tetrahedron*, whilst {1$\bar{1}$1} is termed a *negative tetrahedron*, but this use of positive and negative is open to grave objections, and we shall prefer to distinguish them by quoting their appropriate indices.

If we insert a pole 210 in the stereogram of Fig. 102, it will be seen that the pole 120 is introduced by the operation of a diagonal plane of symmetry, and the form {210} in this class has the shape, the tetrahexahedron, which we have already described in the holosymmetric class. A general pole *h k l* is only repeated twenty-four times, but the form thus generated is, of course, a different one from the general form of the pyrite class. This brief discussion, then, should have helped to make clear that some of the special forms may differ in shape in two different classes, whilst others remain the same, so that no special form is uniquely characteristic of any one particular class—we shall again encounter the pentagonal dodecahedron and the tetrahedron in still other classes of the cubic system. Only the general form has a shape uniquely related to the symmetry of a particular class, and so we shall be justified eventually in naming each of the thirty-two crystal classes by reference to the thirty-two different general forms.

THE TETRAGONAL SYSTEM. (HOLOSYMMETRIC CLASS)

Mercurous chloride, HgCl, can be prepared in crystals which sometimes have the shape of Fig. 104. At first sight, this might be a cubo-

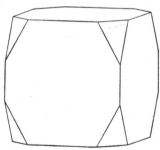

octahedron. The three directions of 'cube' edges can be selected as the directions of crystallographic axes *x y z*, and the 'octahedral' plane satisfies the requirements for a parametral plane 111. Measurement of such a crystal, however, will show that the angles over the edges of each triangular face are not all equal; from 111 on to the front face of the crystal = 49° 5', which is the value also of the angle from 111 to the side face,

FIG. 104. A crystal of mercurous chloride.

but the angle made by 111 with the top face = 67° 50'. The triangular faces are isosceles, not equilateral, and there are no triad axes normal to them; mercurous chloride does not crystallise in the cubic system. The vertical axis is still a tetrad axis, but the x and y directions are diad axes only; the crystal belongs to the tetragonal system. There are other symmetry elements also present, and the complete group is inserted in the stereogram, Fig. 105; there is a vertical tetrad axis, with four planes of symmetry intersecting in it, a horizontal plane of symmetry, four horizontal diad axes (one normal to each of the vertical planes of symmetry) and a centre.

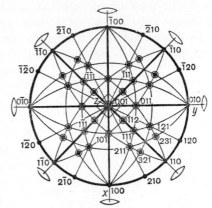

FIG. 105. Stereogram of a holosymmetric tetragonal crystal.

We now proceed to study the number of possible different kinds of special forms, and the nature of the general form, exactly as we did for the holosymmetric class of the cubic system.

The face 100,* normal to the x axis, when repeated to satisfy the symmetry, has associated with it the faces 010, $\bar{1}$00, 0$\bar{1}$0, and no others. There is nothing in the symmetry of this class to demand the presence of the face 001 (nor of 00$\bar{1}$) if the face 100 is given; that is, the form {100} here consists of four vertical faces only, a square-sectioned tube with open ends (Fig. 106). All the forms which we discussed in the cubic system enclosed space—they are *closed forms*, any one of which could be present alone on an actual crystal, and we may have thought that this was a necessary characteristic of any form. Looking back at the definition of a form

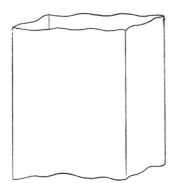

FIG. 106. Tetragonal prism {100}.

* The student should note that this method of inserting indices in the projection as we go along is adopted only for teaching purposes—the lecturer ' knows the answer ', and can insert indices before the projection is complete. In actual practice (p. 95) the crystal is measured completely and a projection is made before it is possible to determine the symmetry and hence even the correct system. Only then can the axes $x\,y\,z$ be chosen, a parametral plane be selected and the indices of the other forms present be determined. In an actual problem, therefore, *never index until the projection is complete.*

(p. 9), it will be seen that this is not so; a form is an ' assemblage ' of faces, but if a particular form does not enclose space it clearly cannot exist by itself on an actual crystal. In {100} of the tetragonal crystal we have encountered for the first time an *open form*. Faces parallel to the vertical axis are called *prism* faces, and {100} is a *tetragonal prism* or *square prism*. The plane 001 is thus not necessarily present on a

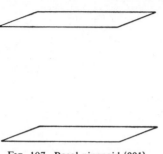

crystal showing the form {100}. It is present, however, in the crystal of Fig. 104, and we insert the corresponding pole in the stereogram; $00\bar{1}$ is then also present (by the operation of the centre of symmetry, for example), and {001} consists of a pair of parallel faces. Such a form, again, is open (Fig. 107), the planes having no defined shape until we know what other forms are present on the crystal. A plane normal to the

FIG. 107. Basal pinacoid {001}.

vertical axis is described as *basal*, and a form composed of a pair of parallel faces is a *pinacoid* (πίναξ, a board), hence {001} is a *basal pinacoid*.

The pole of one of the triangular faces of Fig. 104 lies on the trace of a diagonal plane of symmetry (Fig. 105), since the face itself is normal to the plane of symmetry. In a stereogram of mercurous chloride, the pole is inserted at the correct distance from the centre to correspond to the measured angle of 67° 50′, and is repeated to give eight positions in all—the eight triangular faces of the crystal (Fig. 104) all belong to the form {111}. This somewhat resembles the octahedron of the cubic system, but the faces are isosceles, not equilateral, triangles, and only the horizontal edges form a square. The top and bottom halves are symmetrical, repeated over the horizontal plane of symmetry, and the form is called a *tetragonal bipyramid* (Fig. 108). In the cubic system {111} was generated from a face which occupied a unique position normal to a triad axis, but here

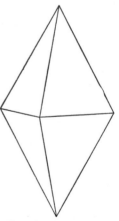

FIG. 108. A tetragonal bipyramid.

there is no such unique position anywhere along the diagonal symmetry plane between the vertical tetrad axis and the horizontal diad axis.

Hence {111} is now typical of a family of tetragonal bipyramids {$h\,h\,l$}, which differ from one another only in being relatively more acute if $h>l$ and more obtuse (Fig. 109) if $h<l$.

Similarly, since there is no diad axis normal to the possible face 101, there is no unique position for an $h\,0\,l$ pole anywhere in the zone between 001 and 100.

All forms {$h\,0\,l$}, including {101}, are members of another family of *tetragonal*

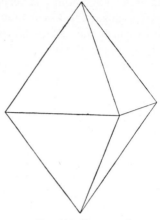

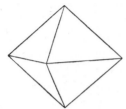

FIG. 109. Another member of the family of tetragonal bipyramids {$h\,h\,l$}.

FIG. 110. Tetragonal bipyramid {$h\,0\,l$}.

bipyramids (Fig. 110) which are exactly the same kind of shape as the {$h\,h\,l$} bipyramids, and differ from them only in presenting a face to the front instead of an edge.

The possible face 110 occupies a unique special position, for it is normal to one of the horizontal diads. To it belong also the faces 1$\bar{1}$0, $\bar{1}$$\bar{1}$0 and $\bar{1}$10, and the form {110} (Fig. 111) is a *tetragonal prism* identical in shape with {100} and differing from it only in the same way as the bipyramids {$h\,h\,l$} differ from the bipyramids {$h\,0\,l$}. This difference has sometimes been termed a difference of *order*, the family {$h\,h\,l$} being described as *tetragonal bipyramids of the first order*, whilst the family {$h\,0\,l$} comprises *tetragonal bipyramids of the second order*. Just as we supposed that cubic crystals were built from a unit of pattern which was itself a cube, so we assume that the underlying unit in a tetragonal crystal is a right square prism the dimensions of which depend upon the particular substance under consideration.

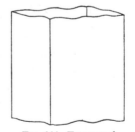

FIG. 111. Tetragonal prism {110}.

The two families of bipyramids therefore bear different relationships to the underlying structure, but the distinction as one of order serves no useful purpose. In Fig. 104 the crystal of mercurous chloride is so orientated that we have naturally taken the diad axes which are normal to the vertical faces as the x and y axes. The other pair of diad axes could equally well be so chosen, the crystal being turned through an angle of 45° about the vertical axis. This alternative choice of axes interchanges the forms of first and second order; we shall therefore distinguish the related forms by quoting the appropriate indices for a given choice of axes.

There remains one further kind of special position, that of a face $h\,k\,0$, parallel to the tetrad axis but not normal to a diad axis. Such

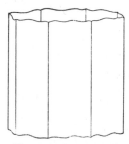

FIG. 112. Ditetragonal prism $\{h\,k\,0\}$.

a prism face is reflected across the vertical planes of symmetry and repeated by the tetrad axis, so that there are eight faces in the form. It is not called an octagonal prism, however; true octagonal symmetry is impossible in crystals, and the significant feature in this kind of form is that the faces meet alternately in more obtuse and more acute edges, four of each kind. We therefore term it a *ditetragonal prism*. It is an open form (Fig. 112).

Any further position for a possible face must be a general one, and it remains to examine the nature of the general form. A pole such as 321, lying off the trace of any element of symmetry in the stereogram (Fig. 105), is repeated eight times above the primitive and symmetrically eight times below. As with the ditetragonal prism, however, the repetition is not regularly eightfold, but in four pairs, and the appropriate name is that of a *ditetragonal bipyramid* (Fig. 113). We have been studying the *ditetragonal bipyramidal* class of the tetragonal system. (Objections have sometimes been raised to the use of the two prefixes *di-* and *bi-*, and some crystallographers write ditetragonal dipyramid. We shall find it useful, however, to use the prefix *bi-* to denote that a form consists of two portions repeated over a plane of symmetry, and the prefix *di-* to denote repetition of pairs of faces around an axis, as explained above.)

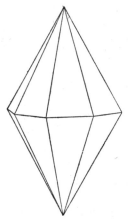

FIG. 113. Ditetragonal bipyramid $\{h\,k\,l\}$.

If the pole 211 is inserted in the stereogram, by drawing the appropriate zones, and repeated to satisfy the symmetry, it will be found that {211}, like {321}, is a ditetragonal bipyramid—a general form. How does it happen that such a ' special '-looking index as 211 belongs to a general form? We must remind ourselves that we have now passed from the very regular cubic system to one in which it is no longer possible to select a parametral plane which makes equal intercepts on all three axes. The assignment of the index {111} to a particular bipyramid {h h l} defines the ratio of the c unit of measurement, parallel to the z axis, to the a unit of measurement parallel to the x and y axes; it has fixed the axial ratio c : a for this description of the crystal. The parametral plane makes on the axes intercepts in the ratio $\frac{a}{1} : \frac{a}{1} : \frac{c}{1}$. A face h h l makes intercepts in the ratios $\frac{a}{h} : \frac{a}{h} : \frac{c}{l}$, and is thus symmetrical to the diad axes which are the x and y axes—it belongs to a special form, a tetragonal bipyramid. A face such as 211, however, makes intercepts in the ratios $\frac{a}{2} : \frac{a}{1} : \frac{c}{1}$, and is thus in no way specially related to the elements of symmetry—it belongs to a general form.

We may tabulate here the various kinds of form which are found in this class of the tetragonal system:

Special forms.   Tetragonal prisms {100}, {110}.
                                Ditetragonal prisms {h k 0}.
                                Basal pinacoid {001}.
                                Tetragonal bipyramids {h h l}, {h 0 l}.
General forms.   Ditetragonal bipyramids {h k l}.

Fig. 114 represents a crystal of tetramethylammonium iodide, $N(CH_3)_4I$. The tetragonal prisms {100} and {110} are clearly both present; in the setting of the figure, {100} is large and {110} small. The larger terminal faces belong to a tetragonal bipyramid {h h l} and could be indexed as {111}. The edges of this bipyramid are modified by small faces of a bipyramid {h 0 l}, which, from its zonal relationships with {111} is clearly the form {101}. No faces of a general form are present, so that we could not be certain from the appearance of this particular crystal that it is correct to assign the substance to the ditetragonal bipyramidal class.

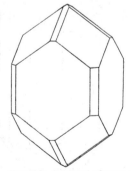

FIG. 114. A crystal of tetra-methylammonium iodide

Fig. 115 illustrates a crystal of mercurous chloride of more complex development than the simple habit of Fig. 104.   The tetragonal prism

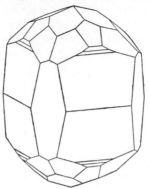

FIG. 115. A crystal of mercurous chloride.

100} is small; there are four different tetragonal bipyramids $\{h\,h\,l\}$ (any one of which might be chosen as the parametral form {111}), one tetragonal bipyramid $\{h\,0\,l\}$ and one ditetragonal bipyramid.

THE ORTHORHOMBIC SYSTEM (HOLOSYMMETRIC CLASS)

(The abbreviation of the name of this system to *Rhombic*, sometimes used, is undesirable.)

In Fig. 116 is illustrated a crystal of lead sulphate, $PbSO_4$.   The rectangular shape of certain of the faces suggests the presence of diad axes, and measurement of such a crystal would show that these faces are mutually orthogonal.   The only other symmetry elements are three

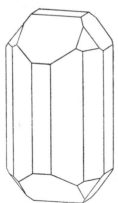

FIG. 116. A crystal of lead sulphate.

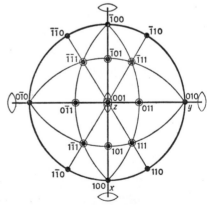

FIG. 117. Stereogram of a holosymmetric orthorhombic crystal.

planes of symmetry (one normal to each diad axis) and a centre. This symmetry is marked on the stereogram, Fig. 117. We shall naturally choose the directions of the diad axes as the crystallographic axes *x y z*. There are three unique special positions, normal to each of the three diad axes. A face such as 100 is repeated only once, in the position 100, to give a *pinacoid* {100}; similarly, {010} and {001} are two further possible pinacoids, the latter often called a *basal pinacoid* by analogy with the use of this name in the tetragonal system.

In the zone between 100 and 010 will lie possible faces *h k* 0, of which 110 is the simplest, and these will give rise to a family of special forms since they occupy a special kind of position normal to a plane of symmetry (and therefore also parallel to a diad axis). Since the crystal is orthorhombic and not tetragonal in symmetry, however, we must suppose that the underlying unit of pattern is a rectangular parallelepiped (a brick-shaped unit) and not a right square prism. Hence even the face 110 will not lie at 45° between 100 and 010, but will make unequal intercepts on the *x* and *y* axes (Fig. 117). The face is repeated four times by the symmetry elements —110, 1Ī0, ĪĪ0, Ī10—and the form {110} is an open prism form (Fig. 118). Unlike {110} in the

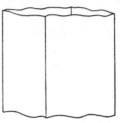

FIG. 118. Orthorhombic prism {*h k* 0}.

tetragonal system, the right cross-section is now a rhombus and not a square, and all the forms {*h k* 0} make up a family of *orthorhombic prisms*.

A face {0 *k l*} is likewise repeated four times to give a similar open

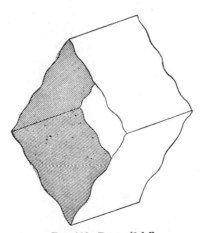

FIG. 119. Dome {0 *k l*}.

form (Fig. 119) with a rhombus cross-section, but differently situated in space. Such forms {0 k l} are called *domes* from their resemblance to the gable-roof of a house (δῶμα). Forms {h 0 l} comprise a family of similar domes {Fig. 120) with their edges parallel to the *y* axis.

We have now exhausted the possible kinds of special position, and any other possible plane h k l must be generally related to the elements of symmetry—the systematic triangle (p. 57) is now a quadrant of the stereogram. It may be surprising at first to realise that even {111} is thus a general form, but we must remember that in a structure built

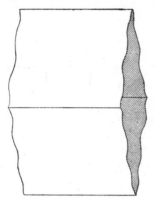

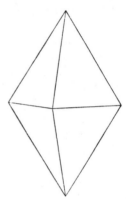

FIG. 120. Dome {h 0 l}.

FIG. 121. Orthorhombic bipyramid { h k l}.

from brick-shaped units it is impossible to choose a parametral plane making equal intercepts even on one pair of axes; the axial units *a*, *b*, *c* are now perforce all unequal (in other words, neither of the *axial ratios a/b, c/b* can be unity in the orthorhombic system). Thus the plane 111 makes intercepts $a/1$, $b/1$, $c/1$ on the three axes; these are unequal and unrelated, and the plane is generally situated. Any such plane is repeated by reflection to give four faces above the primitive (Fig. 117) and four faces symmetrically below, and belongs to a general form of the *orthorhombic bipyramidal class* (Fig. 121).

From the customary list of forms, it will be seen that as the symmetry falls from system to system this list is gradually growing simpler:

Special forms.   Pinacoids {100}, {010}, and (basal) {001}.
Prisms {h k 0} and domes {h 0 l}, {0 k l}.
General forms.   Orthorhombic bipyramids {h k l}.

It was formerly a convention that an orthorhombic crystal should be so orientated before description that it was possible to choose a parametral plane giving $a/b<1$. The x axis, along which the shorter units of measurement were employed, was then called the *brachy-axis* and the y axis the *macro-axis* ($\beta\rho\alpha\chi\acute{v}s$, short; $\mu\alpha\kappa\rho\acute{o}s$, long). This was sometimes also symbolised $\breve{a}$, $\bar{b}$, $\dot{c}$. Forms such as {100} and {h 0 l}, developed from faces parallel to the macro-axis, were then called *macropinacoid* and *macrodomes* respectively, whilst {010} was the *brachypinacoid* and {0 k l} *brachydomes*. It is now recognised that analogies between the structural plans of related substances are sometimes better expressed by an orientation in which $a/b>1$ for a particular substance, and once this convention is not rigidly observed the corresponding nomenclature is almost valueless. We shall not use it here, but shall distinguish the forms by their appropriate indices just as we distinguish ' order ' in the tetragonal system. It is clear, for example, that the pinacoid {010} consists of a pair of faces parallel to the direction which we have chosen as the x axis, quite independently of the relative values of a and b units. Finally, it may be observed that the three diad axes are all of similar crystallographic significance and any one might be selected as the z direction, so that there are in all six possible ways of orientating every orthorhombic crystal. There is no real difference ultimately between the ' basal ' pinacoid and the pinacoids {100} and {010}; nor between prisms and domes. For this reason we have listed them as equivalent above, and some authors talk of ' domal prisms ', but once a particular diad has been selected as the z direction it is convenient to retain the term *prism* for forms {h k 0} parallel to it.

Fig. 122 represents a crystal of caesium perchlorate, $CsClO_4$, of somewhat tabular habit. The three pinacoids {001}, {010} and {100} are

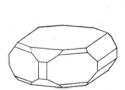

FIG. 122. A crystal of caesium perchlorate.

FIG. 123. A crystal of sulphur.

developed, the last very small. There are present also one prism one dome {h 0 l}, one dome {0 k l} and one orthorhombic bipyramid. In Fig. 123 is represented a crystal of orthorhombic sulphur of simple habit. (Sulphur is *polymorphous*, or *allotropic*, and also crystallises,

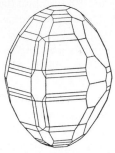

under different conditions, in crystals not belonging to this system. The substance we are considering here is the modification which is formed also under natural conditions as a mineral.) The basal pinacoid {001}, a dome {0 *k l*} and two bipyramids are developed. Fig. 124 represents a much more highly modified crystal of the same substance; the three pinacoids are all present, and there are five domes, a prism and fourteen examples of the general form.

FIG. 124. A highly-modi-
fied crystal of sulphur.

THE MONOCLINIC SYSTEM (HOLOSYMMETRIC CLASS)

Fig. 125 illustrates a crystal of borax, $Na_2B_4O_7 . 10H_2O$. There is a number of vertical ' prism ' edges, indicating an obvious choice for the z direction. There are edges normal to this, running right and left, which can be chosen to determine the y direction as in the previous systems. The right-hand vertical face, however, shows no direction of edge normal to the z direction, and examination of a number of crystals of borax would convince us that such a direction is not a ' possible edge '. We must suppose that borax crystals are built up from a unit of pattern which is not an orthogonal prism (Fig. 126), and acting in conformity with the formulation of the Law of Rational Indices it becomes impossible to describe such a

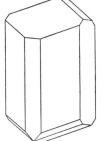

FIG. 125. A crystal
of borax.

crystal in terms of a set of orthogonal axes. The y and z directions can still be chosen normal to each other, but the x direction, whilst lying in a plane through the z axis normal to the y z plane, slopes at an angle to the z direction which is not 90° (Fig. 126); hence the name of the system—*monoclinic*, one axis inclined.

The figure of a borax crystal shows clearly the lack of a horizontal plane of symmetry, but there is still a vertical plane of symmetry (parallel to the plane of the x z axes which we have chosen) and a diad axis normal to this (parallel to the y direction); the only other element of symmetry present is a centre. It is now a universally-accepted convention* that in a monoclinic crystal the diad axis shall be chosen as the y direction; conventionally, also, the x direction slopes *downwards*

* See, however, p. 272.

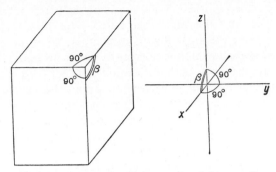

FIG. 126. Structural unit of a monoclinic crystal and the crystallographic axes derived from it.

towards the observer, so that the angle β between $+x$ and $+z$ *is always obtuse*. There has been some confusion over this latter point, however, and we shall refer to it again shortly.

The non-orthogonal character of the axes opens up a new difficulty in making a stereographic projection; if the z direction is still placed normal to the paper, as hitherto, the y direction still runs right and left in the plane of the paper, but $+x$ emerges in front *below* the plane of the paper whilst $-x$ projects at the back above this plane (Fig. 127). This is the method of projection most widely used at the present time, but for some purposes it is better to make the projection in a different

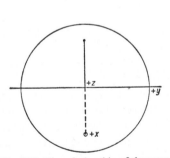

FIG. 127. The relationship of the crystallographic axes of a monoclinic crystal to a stereographic projection on a plane normal to the z-axis. (Formally, the axes should be represented only by their poles, since the stereogram is a projection of the *surface* of the sphere; it is often convenient in practice, however, to insert the axes as they would be seen inside a transparent sphere.)

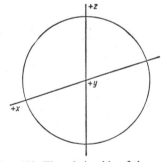

FIG. 128. The relationship of the crystallographic axes to a stereographic projection on a plane normal to the y-axis.

attitude. The plane of the paper is passed through the spherical projection normal to the y axis instead of to the z axis; the projecting point *P* is situated on the −*y* axis, so that the plane of projection is parallel to the crystallographic plane 010 (Fig. 128). The *x* and *z* directions, making the angle *β* with each other, lie in the plane of the paper, whilst the *y* direction is normal to the projection. The student should practise the use of both orientations, but for the present we shall follow the former method, projecting on a plane normal to the *z* axis.

In such a projection, the vertical faces of Fig. 126, meeting in edges parallel to the *z* axis, must project as poles lying on the primitive circle. The face 100 is parallel also to the *y* axis, and so its pole can be inserted in the stereogram (Fig. 129); it is repeated at $\bar{1}00$, so that the form {100} is a pinacoid. The face 010 is normal to the *y* axis, and with the parallel face $0\bar{1}0$ constitutes another pinacoid. The face 001

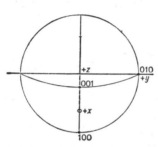

Fig. 129. The poles of the planes 100, 010 and 001 inserted in the projection in Fig. 127.

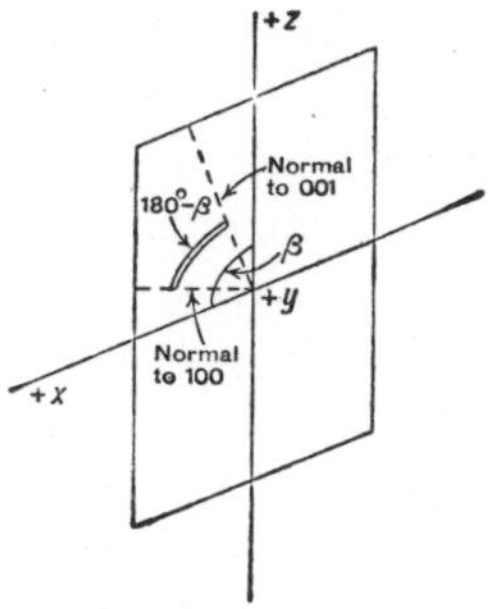

Fig. 130. Plan on the *xz* plane, showing the relationship of the axial angle *β* to the interfacial angle 100∧001.

is parallel to the *y* axis and also to the direction selected as the *x* axis; it is therefore not normal to the *z* axis, and its pole no longer lies at the centre of the stereogram, but at an angular distance from the centre corresponding to an inclination forwards equal in amount to the inclination of the *x* axis below the horizontal. This relationship is most clearly seen in a plan of the crystal on the *xz* plane (Fig. 130); since the *x* axis is parallel to 001 and the *z* axis is parallel to 100, it is clear from this figure that the normal crystallographic angle 100∧001 = 180° − *β*, where *β* is the *obtuse* angle between +*x* and +*z*. This normal crystallographic angle can be measured directly on the crystal, and the pole 001 is then readily inserted in the projection (Fig. 129). It, also, belongs to a pinacoid, {001}, a basal pinacoid by analogy with the preceding systems.

We noted above the two conventions applying to the orientation of a monoclinic crystal—the diad axis is always selected as the *y* axis, and the +*x* axis always slopes downwards towards the observer. The angle β, defined as the angle between $+x$ and $+z$, must therefore be obtuse, and the crystallographic angle $100 \frown 001$ must be acute. In accordance with this, we find for example in Groth's *Chemische Krystallographie* that in lead chromate, $PbCrO_4$, $\beta = 102° 27'$. This substance occurs as a mineral, crocoite, and if we consult a mineralogical reference book we shall probably find the entry $\beta = 77° 33'$. Such statements seem to have arisen from an impression that an acute value of an angle was more easily understood and manipulated than was an obtuse value; in some books they are partially justified by defining β as the angle between $-x$ and $+z$, but this subterfuge only adds to the confusion, as in the next system, the triclinic, the definition cannot be retained. This perverse habit is slowly being abandoned in current texts, and the student should cultivate the habit of quoting an acute value if he refers to the crystallographic angle $100 \frown 001$ and the corresponding obtuse value if he refers to the axial angle β. Meanwhile he will keep clear of this awkward trap if he gets firmly fixed in his mind that crystallographers *never orientate a monoclinic crystal with $+x$ projecting upwards in front.*

Returning now to a discussion of the possible forms in this symmetry group, we consider next the repetition of a face $h\,k\,0$. Such a face is reflected across the plane of symmetry to give $h\,\bar{k}\,0$, and the diad axis (or the centre) adds two further faces to give the form $\{h\,k\,0\}$, an open *prism.* Strictly speaking, this is a general form; the only possible special positions in this group are the unique position of 010 normal to the diad axis, and the family of positions $h\,0\,l$ (of which 100 and 001, considered above, are particular examples) parallel to the diad axis. This class is therefore called the *prismatic class* of the monoclinic system. The list of possible forms is now quite brief:

Special forms. Pinacoid $\{010\}$.
                Pinacoids $\{h\,0\,l\}$.
General forms. Prisms $\{h\,k\,l\}$.

It is customary in practice, however, to depart from this somewhat rigid position and to name various forms differently if they bear a different relationship to the *crystallographic* axes. (Compare the distinctions of ' order ', p. 65.) We began to do this above, when we suggested calling $\{001\}$ a *basal pinacoid,* and though this irregularity of nomenclature may seem undesirable at first sight, it is extremely convenient in practice in view of the close relationships which often exist between certain orthorhombic and allied monoclinic crystals.

The y axis is normal to the z axis, and may be called the *ortho-axis,* whilst the inclined *x* axis is termed the *clino-axis*. {100} is then distinguished as the *orthopinacoid*, and {010} as the *clinopinacoid*; {001}, though not normal to the *z* axis, is, by analogy with the orthorhombic system, the *basal pinacoid*. Forms {0 *k l*} can be called *clinodomes*, and forms {*h* 0 *l*} *orthodomes*; the latter, strictly speaking, are *hemi-domes*, since the presence of *h* 0 *l* does not imply the necessary presence of *h̄* 0 *l*, and each form consists only of two faces. Finally, {*h k l*} forms are by analogy *hemi-pyramids*, but we must recognise the informality of this nomenclature by keeping to the correct name for the class, the prismatic class. The relationship between the two kinds of nomenclature may be tabulated:

Clinopinacoid {010}.   Clinopinacoid {010}.
Orthopinacoid {100}.   ⎫
Basal pinacoid {001}.   ⎬   Pinacoids {*h* 0 *l*}.
Hemi-orthodomes {*h* 0 *l*}. ⎭
Clinodomes {0 *k l*}.   ⎫
Prisms {*h k* 0}.   ⎬   Prisms {*h k l*}.
Hemi-pyramids {*h k l*}.   ⎭

Fig. 131 illustrates a common habit of hydrated calcium sulphate, $CaSO_4 . 2H_2O$, the mineral gypsum. The clinopinacoid {010} is well-developed and is accompanied by two examples of the general form. The simplest indices which we could allocate to these would be {110}

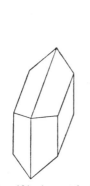
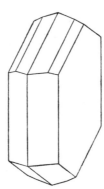

and {011}, determining the *x* and *z* directions parallel to the edges bounding the face 010, though in the conventional description of gypsum this has not been done. The crystal of sodium bicarbonate, $NaHCO_3$, depicted in Fig. 132, is rather pronouncedly tabular parallel to the clino-pinacoid, which is here accompanied by four examples of the general form. The conventional indexing describes

FIG. 131. A crystal of gypsum.   FIG. 132. A crystal of sodium bicarbonate.

one of these as the 'prism' {110}, and assigns the indices {111}, {121} and {11$\bar{1}$} to the remaining 'hemi-pyramids'; the value of β is 93° 19′ in this description (notice the pseudo-orthorhombic aspect of Fig. 132), corresponding to the choice of a possible edge, and not an actual edge, of the crystal of Fig. 132 as the x axis. The crystal of trona, $Na_2CO_3 . NaHCO_3 . 2H_2O$, Fig. 133, illustrates a different kind of habit, elongation parallel to the y axis resulting from a prominent development of the pinacoids (including 'hemi-orthodomes') $\{h\,0\,l\}$, of which four examples are present; in addition, there are three examples of the general form.

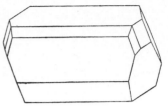

Fig. 133. A crystal of trona.

THE TRICLINIC SYSTEM (HOLOSYMMETRIC CLASS)

Fig. 134 illustrates a crystal of hydrated ferrous sulphate, $FeSO_4 . 5H_2O$ (not to be confused with 'green vitriol', $FeSO_4 . 7H_2O$, which is monoclinic). Measurement of such a crystal would fail to reveal any

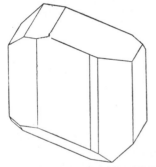

Fig. 134. A crystal of $FeSO_4 . 5H_2O$.

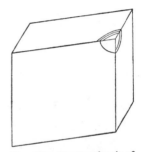

Fig. 135. Structural unit of triclinic crystal.

orthogonal relationships; there are no mutually perpendicular faces or edges, and we must conclude that the underlying unit of pattern is a non-orthogonal parallelepiped (Fig. 135); any suitable choice of prominent directions of edge will result in a set of non-orthogonal axes (Fig. 136), in which no one of the axial angles α, β, γ is 90° (hence *triclinic*, all three axes inclined). There are no axes of symmetry and no planes of symmetry; the only remaining symmetry element is a centre.

In constructing a stereogram nothing is gained by any departure from the usual convention that the z axis shall be normal to the plane of projection. Neither the x direction nor the y direction will then

lie in the plane of the projection; as in the monoclinic system, the angle
β between $+x$ and $+z$ is conventionally obtuse, so that $+x$ projects

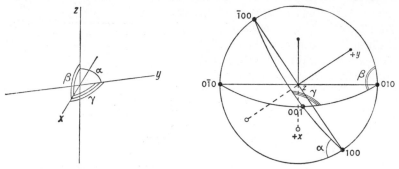

FIG. 136. Crystallographic axes of a triclinic crystal.   FIG. 137. Stereogram of a triclinic crystal.

below the paper in front as before (Fig. 137) whilst $+y$ projects above
or below the paper according to the acute or obtuse value of the angle
α.   Since there are no symmetry elements to guide the conventional
choice of even one axial direction, it will be readily seen that different
observers may make widely differing selections in different descriptions
of the same substance; attempts have been made from time to time to
introduce conventional rules, but it cannot be said that any are widely
accepted at present, except perhaps that α, as well as β, shall be obtuse.

The directions of the x and y axes cannot be inserted in the stereogram in actual practice until the planes 100, 010 and 001 have been
projected.   Since 100 and 010 are parallel to the z axis, their poles lie
on the primitive.   If the xz axial plane is still set north-south on the
projection, 010 occupies its usual position on the primitive (Fig. 137).
Measurement of the crystallographic angle 010$\frown$100 enables us to
insert 100, and 001 must be located by the intersection of two small
circles of radii corresponding to the measured crystallographic angles
100$\frown$001 and 010$\frown$001.   Since the x axis is parallel both to 010 and
to 001, it is the zone-axis of the zone through these two faces, and its
point of emergence can be located as the pole of the corresponding
great circle; similarly, the y axis emerges at the pole of the great circle
through 100 and 001 (Fig. 137).   This relationship between the axial
directions and the associated great circles enables us to mark in the
projection the angles which correspond to the values of the axial
angles α, β, γ.   The angle α, for example, is the angle between $+y$ and
$+z$; but y is normal to the zone $100 - 001$ and z is normal to the zone
$100 - 010$, so that α is given by the angle between these two zones.

Similarly, we can indicate the values of β and γ, and it is clear at once that there is an important difference here from the corresponding aspect of the monoclinic system; whereas in the monoclinic crystal the crystallographic angle 100⌒001 is simply related to the axial angle β ($\beta = 180° - 100⌒001$), there is no such direct relationship in a triclinic crystal between the measured interfacial angles 100⌒010, 010⌒001, 001⌒100, and the axial angles α, β, γ.

The discussion of possible forms now becomes very brief. Since no special relationship to a centre is possible, the distinction of special forms from general forms has disappeared in this class. Every form consists of a pair of parallel faces, a pinacoid; the class is the *pinacoidal class* of the triclinic system. Once the crystallographic axes have been chosen, it is occasionally convenient to carry over here, also, the nomenclature of the orthorhombic system and to speak of domes and pyramids in relation to this particular set of axes, but if this departure from strict nomenclature is tolerated it must be constantly borne in mind that each form consists only of a pair of parallel faces. Some typical triclinic pinacoidal crystals will be illustrated later, when we return to study the triclinic system in greater detail; an illustration would add nothing at this stage, since we have seen that every form present must be a pinacoid.

THE HEXAGONAL SYSTEM (HOLOSYMMETRIC CLASS)

A crystal such as Fig. 138 portrays clearly possesses a hexad axis, and so must be allocated to the hexagonal system. The simplest assumption we can make about the unit of pattern of such a crystal is

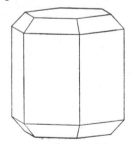

FIG. 138. A crystal showing a hexad axis.

FIG. 139. Structural units of a hexagonal crystal.

that it has the shape of one of the right prisms based on a 60° rhombus shown in Fig. 139. The vertical prism edges of such a unit (and therefore of the prism form of Fig. 138) will clearly be chosen as the direction of a vertical z axis. Normal to this direction are the horizontal edges of the prism units of Fig. 139; two of these edges can therefore be used to determine x and y directions normal to z. The x and y

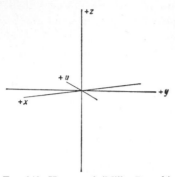

axes thus make an angle of 120° with each other, but which of the three possible directions shall be selected as the x direction and which as y? The three directions are all of equivalent significance, and so it is customary in this system to use all three; hexagonal crystals are thus described in terms of four crystallographic axes (Fig. 140), three horizontal axes x, y and u, at an angle of 120° to each other, normal to the vertical z axis. The Millerian index notation which we have used in the

FIG. 140. Hexagonal (Miller-Bravais) crystallographic axes.

preceding crystal systems was adapted to this set of crystallographic axes in the hexagonal system by A. Bravais,* and the indices which we shall use are therefore known as *Miller-Bravais Indices*. They are written to refer to the axes in the order $x\,y\,u\,z$, and each index symbol thus contains four figures —e.g. $10\bar{1}1$, $11\bar{2}2$, or generally $h\,k\,i\,l$. (Some authors write this general symbol $h\,i\,k\,l$, or even $h\,k\,l\,i$, causing unnecessary confusion, but the order of reference to the axes is always that which we have described.) It will be clear that the first three indices are not independent, since they refer to the intercepts on three fixed coplanar axes; the student can soon prove for himself that $h+k+i=0$. The reason for the

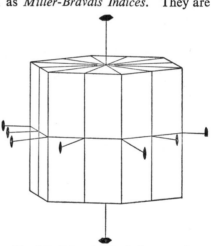

FIG. 141. Prism showing holosymmetric hexagonal symmetry.

use of a redundant index in this way will be more easily understood when we have had some practice in the Miller-Bravais notation (see p. 83).

---

* Auguste Bravais was born at Annonay, southern France, in 1811, and became a naval officer. His interests in astronomy and other branches of mathematical physics ultimately led him to a Chair of Physics in Paris, but he is best known to crystallographers for his work on the theory of crystal structures, to which we shall refer later. He died at Versailles in 1863.

The symmetry of the holosymmetric class of the hexagonal system is illustrated by the hexagonal prism (Fig. 141). There are six vertical planes of symmetry intersecting in the hexad axis, six horizontal diad axes, a horizontal plane of symmetry and a centre. This symmetry is marked on the stereogram (Fig. 142), where the conventional orientation of the axes is also illustrated. The diad axis chosen as the y direction runs horizontally to the right, as before, so that the $+x$ direction now runs downwards towards the left and the $+u$ direction upwards towards the left. The z direction is of course normal to the paper, the plane of projection.

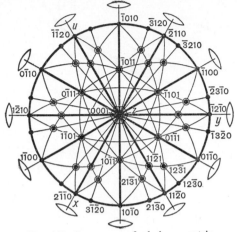

FIG. 142. Stereogram of a holosymmetric hexagonal crystal.

To begin a study of the possible kinds of form, we can insert the basal plane, the index of which must be 0001. It is repeated in 000$\bar{1}$ to give the *basal pinacoid* {0001}. As in the earlier systems discussed above, this is an open form with no defined shape until it is known what other forms are also present. The prism, the faces of which intersect the basal pinacoid in the chosen $x\,y\,u$ directions, may now be inserted. These faces are parallel to the z axis, and their poles therefore lie on the primitive; each pair of faces is parallel to one of these horizontal axes. Each face makes equal intercepts, of opposite sign, on the two horizontal axes which it intersects; the index of the form is therefore $\{h\,0\,\bar{h}\,0\}$, or in simplest terms $\{10\bar{1}0\}$. The six faces necessitated by the symmetry make up an open *hexagonal prism* (Fig. 143).

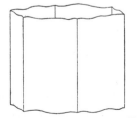

FIG. 143. Hexagonal prism {10$\bar{1}$0}.

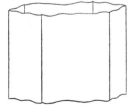

FIG. 144. Hexagonal prism {11$\bar{2}$0}.

Truncating the edges of this prism are the faces of a second possible prism, cutting two horizontal axes symmetrically; its index is $\{h\,h\,\overline{2h}\,0\}$, i.e. $\{11\overline{2}0\}$. It is a further hexagonal prism (Fig. 144). Any further

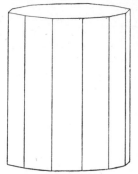

possible prism face $h\,k\,i\,0$, such as $21\overline{3}0$ (bear in mind that $h+k+i=0$), is not normal to a vertical plane of symmetry and is therefore repeated twelve times altogether, in six pairs; the form $\{h\,k\,i\,0\}$ is a *dihexagonal prism*. The form $\{21\overline{3}0\}$ is illustrated in Fig. 145, closed by the basal pinacoid $\{0001\}$.

A possible face in the zone 0001–$10\overline{1}0$ will have indices $h\,0\,\overline{h}\,l$, of which the simplest example is $10\overline{1}1$. Such a face is repeated six times around the vertical hexad axis and six

FIG. 145. Dihexagonal prism $\{21\overline{3}0\}$ with basal pinacoid $\{0001\}$.

times symmetrically below; the forms $\{h\,0\,\overline{h}\,l\}$ constitute a family of *hexagonal bipyramids*
(Fig. 146). The particular slope of the face $10\overline{1}1$, of course, depends upon the dimensions of the unit of pattern and therefore upon the particular substance under consideration. The plane $10\overline{1}1$ may be considered as the parametral plane of this system (an index 1111 is, clearly, impossible), since its slope defines the ratio of the a units of measurement along the x, y and u axes to the c unit of measurement along the z axis, determining an *axial ratio* c/a of similar significance to that of a tetragonal crystal.

One further special position remains to be considered. Just as $\{h\,0\,\overline{h}\,l\}$ bipyramids modify the edges between the prism $\{10\overline{1}0\}$ and the basal pinacoid $\{0001\}$, so there is also a family of forms $\{h\,h\,\overline{2h}\,l\}$, such as $\{11\overline{2}1\}$, in the zones between the faces of the prism $\{11\overline{2}0\}$ and the pinacoid $\{0001\}$. These are a further family of hexagonal bipyramids, geometrically similar to the forms $\{h\,0\,\overline{h}\,l\}$, but presenting an edge towards

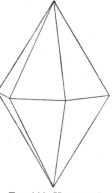

FIG. 146. Hexagonal bipyramid $\{h\,0\,\overline{h}\,l\}$.

us where the latter present a face (Fig. 147). The distinction between these two families of bipyramids, and between the hexagonal prisms $\{10\overline{1}0\}$ and $\{11\overline{2}0\}$, is sometimes described as one of order, but similar objections apply here to those which we advanced against this method of distinction in the tetragonal system (p. 65), and we shall continue to make the distinction by quoting the appropriate indices.

No other specially related position of a face is possible. A face $h\,k\,i\,l$, such as $21\bar{3}1$, is generally related (its pole lies within the systematic triangle) and is part of a general form. Such a face is repeated

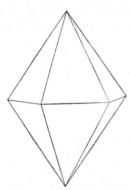

FIG. 147. Hexagonal bipyramid $\{h\,h\,\overline{2h}\,l\}$.

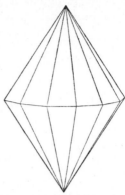

FIG. 148. Dihexagonal bipyramid $\{h\,k\,i\,l\}$.

(Fig. 142) to give six pairs of faces above the plane of projection and six pairs symmetrically below (Fig. 148); we have been studying the *dihexagonal bipyramidal class* of the hexagonal system.

The complete list of forms in this class now reads:

Special forms.   Basal pinacoid $\{0001\}$.

Hexagonal prisms $\{10\bar{1}0\}$, $\{11\bar{2}0\}$.

Dihexagonal prisms $\{h\,k\,i\,0\}$.

Hexagonal bipyramids $\{h\,0\,\bar{h}\,l\}$, $\{h\,h\,\overline{2h}\,l\}$.

General forms.   Dihexagonal bipyramids $\{h\,k\,i\,l\}$.

It is now easy to see, from a consideration of the indices of any one of these forms, why the redundant u index is not omitted altogether. Below are set down the indices of the six faces of the hexagonal prism $\{11\bar{2}0\}$, and the three-figure version which would be obtained by omission of the u index:

$11\bar{2}0$	110
$\bar{1}2\bar{1}0$	$\bar{1}20$
$\bar{2}110$	$\bar{2}10$
$\bar{1}\bar{1}20$	$\bar{1}\bar{1}0$
$1\bar{2}10$	$1\bar{2}0$
$2\bar{1}\bar{1}0$	$2\bar{1}0$

It is easy to see the connection between the six four-figure indices, but this connection is quite obscured in the curious collection in the right-hand column. Omission of either the x or the y figure would secure no more symmetrical result, and so the three are always retained in morphological descriptions. (If, however, the emphasis is placed on the *kind* of plane and it is not important to distinguish which particular face of a form is concerned, as in many problems dealing with the internal structure, a dot or an asterisk may be used in place of the third figure. Thus 11.0 planes are structural planes parallel to any face of the prism $\{11\bar{2}0\}$.)

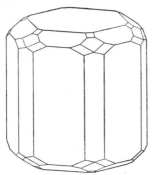

Fig. 149 illustrates a crystal of beryllium aluminium silicate, $Be_3Al_2(Si_6O_{18})$, the mineral beryl, of typical prismatic habit. In the setting shown, the prism $\{10\bar{1}0\}$ is large and the prism $\{11\bar{2}0\}$ smaller. Between these prisms and the basal pinacoid $\{0001\}$ are one example of a hexagonal bipyramid $\{h\,0\,\bar{h}\,l\}$ and two of the bipyramids $\{h\,h\,\overline{2h}\,l\}$. There is also one

FIG. 149. A crystal of beryl.

dihexagonal bipyramid present. If we assign to the bipyramid $\{h\,0\,\bar{h}\,l\}$ the simplest possible index $\{10\bar{1}1\}$ the evident zonal relationships indicate that the other hexagonal bipyramids are $\{11\bar{2}1\}$ and $\{11\bar{2}2\}$, whilst the general form is $\{21\bar{3}1\}$.

THE TRIGONAL SYSTEM (HOLOSYMMETRIC CLASS)

Fig. 150 represents a simple crystal of calcite, $CaCO_3$. A first glance suggests a connection with the hexagonal system, but examination of the terminal faces reveals that the vertical axis is a triad axis only—the system is trigonal and not hexagonal. There is a centre of symmetry but no horizontal symmetry plane. Three vertical planes of symmetry meet in the triad axis, and normal to these three planes are three diad axes which we shall select as the $x\,y\,u$ directions of a Miller-Bravais axial scheme, the z direction being parallel to the triad axis.

Beginning a study of the possible kinds of form, $\{0001\}$ is a *basal pinacoid* as before. By inserting the poles in a stereogram (Fig. 151) and repeating them to satisfy the symmetry, the student can convince himself that $\{10\bar{1}0\}$

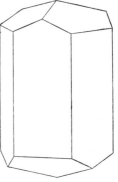

FIG. 150. A crystal of calcite.

and $\{11\bar{2}0\}$ are *hexagonal prisms*, $\{h\,k\,i\,0\}$ a family of *dihexagonal prisms*, and $\{h\,h\,\overline{2h}\,l\}$ a family of *hexagonal bipyramids* exactly as in the

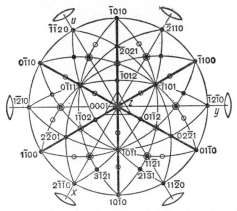

FIG. 151. Stereogram of a holosymmetric trigonal crystal.

hexagonal system. We must continue to use the term *hexagonal* for these forms, though they are special forms of a class of the *trigonal* system, since their faces are disposed regularly every sixty degrees around the triad axis. This reappearance of a number of special forms emphasises the close connection between the trigonal and hexagonal systems, and is one of the reasons why some crystallographers prefer to group all the classes of these two systems together in sub-divisions of one large hexagonal system.

A new feature arises, however, when we consider the repetition of a face $h\,0\,\bar{h}\,l$, such as $10\bar{1}1$, normal to one of the vertical planes of symmetry. The operation of the triad axis gives only three such faces, $10\bar{1}1$, $\bar{1}101$ and $0\bar{1}11$ on top, and the operation of the centre (or of the horizontal diads) gives three parallel faces below. There are no faces of this form symmetrically below the upper faces (see the separate stereogram (Fig. 152), and the form $\{10\bar{1}1\}$ is not a bipyramid but a *rhombohedron* (Fig. 153). The forms $\{h\,0\,\bar{h}\,l\}$ constitute a family of rhombo-

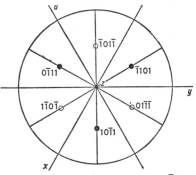

FIG. 152. Stereogram of the form $\{10\bar{1}1\}$ of a holosymmetric trigonal crystal.

hedra which become more and more acute (Fig. 154) as the ratio $h : l$ becomes larger. Though the face $01\bar{1}1$ is not a part of the form

$\{10\bar{1}1\}$ it is a possible face on such a crystal, and its relationship to the symmetry elements shows that it, also, is repeated to give a rhombohedron (Fig. 155) geometrically similar to $\{10\bar{1}1\}$, but presenting an edge towards the observer where the latter presents a face. Developed equally together on a crystal, these two

FIG. 153. Rhombohedron $\{10\bar{1}1\}$.

rhombohedra would simulate a hexagonal bipyramid, but even then each form would retain its own particular characteristics; in calcite, for example, $\{10\bar{1}1\}$ planes are directions of perfect cleavage.

Various nomenclatorial devices have been introduced to differentiate a rhombohedron such as $\{10\bar{1}1\}$ from its geometrically similar 'complementary' rhombohedron $\{01\bar{1}1\}$. Thus, one has been called a *positive rhombohedron* (with an upper face towards the observer) and the other a *negative rhombohedron*, but this mode of distinction seems specially undesirable in view of the established usage of $+$ and $-$ in optical work. *Direct* and *inverse* or *obverse* and *reverse* are more satisfactory terms, but not widely used. We shall distinguish the rhombohedra by their indices; with the conventional setting of the axes, rhombohedra $\{h\,0\,\bar{h}\,l\}$ clearly present an upper face towards the observer (Figs. 153, 154), whilst rhombohedra $\{0\,k\,\bar{k}\,l\}$ present an edge in this position (Figs. 155, 156, 157). Bearing in mind the observed simple zonal relationships of common crystal faces, we shall expect that the rhombohedra most frequently found developed together will be groups such as $\{02\bar{2}1\}$, $\{10\bar{1}1\}$ and $\{01\bar{1}2\}$

FIG. 154. An acute rhombohedron $\{h\,0\,\bar{h}\,l\}$.

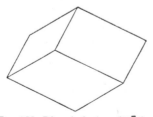

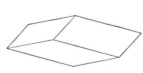

FIG. 155. Rhombohedron $\{01\bar{1}1\}$.   FIG. 156. An obtuse rhombohedron $\{0\,k\,\bar{k}\,l\}$.

developed by successive truncation of polar edges. (The polar edges of a rhombohedron (p. 6) are those which intersect the triad axis.)

No other kind of special relationship to the symmetry elements is possible, and a face such as $21\bar{3}1$ must belong to a general form. Such a face is reflected across a plane of symmetry to give $3\bar{1}2\bar{1}$, and this

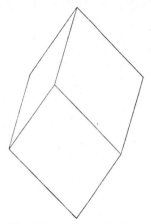

Fig. 157. An acute rhombohedron $\{0\,k\,\bar{k}\,l\}$.

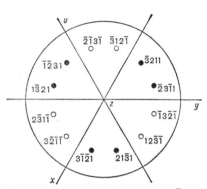

Fig. 158. Stereogram of the form $\{21\bar{3}1\}$ of a holosymmetric trigonal crystal.

pair of faces is repeated three times around the triad axis. The edges between these six upper faces, however, are not now all alike; it is a question of alternate like and unlike edges, not of six similar edges. The form is *ditrigonal* and not hexagonal. Moreover, there is no horizontal plane of symmetry, or any other element of symmetry which operates to give six faces symmetrically below; the upper faces are repeated by the centre (or by the horizontal diad axes) as shown in Fig. 158, and the form $\{21\bar{3}1\}$ is clearly not a bipyramid. Each face of the form is a scalene triangle (Fig. 159), and so the form is called a *ditrigonal scalenohedron*. The name scalenohedron is sometimes used for forms in other systems, so that we should strictly always use the full name for this particular form; when the system under discussion is clear, however, the adjective ditrigonal is often omitted. We have been considering the *scalenohedral class* of the trigonal system.

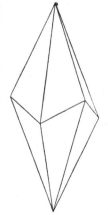

Fig. 159. Ditrigonal scalenohedron $\{h\,k\,i\,l\}$.

The list of forms for this class of symmetry now reads:

Special forms.   Basal pinacoid {0001}.

Hexagonal prisms {10$\bar{1}$0}, {11$\bar{2}$0}.

Dihexagonal prisms {$h\,k\,i\,0$}.

Rhombohedra {$h\,0\,\bar{h}\,l$}, {$0\,k\,\bar{k}\,l$}.

Hexagonal bipyramids {$h\,h\,\overline{2h}\,l$}.

General forms.   Ditrigonal scalenohedra {$h\,k\,i\,l$}.

Fig. 160 depicts a crystal of corundum, Al_2O_3, of tabular habit due to the prominent development of the basal pinacoid {0001}. The other predominant form is a rhombohedron {$h\,0\,\bar{h}\,l$}; in addition, there are small faces of the prism {11$\bar{2}$0}, a rhombohedron {$0\,k\,\bar{k}\,l$} and a hexagonal bipyramid. If the predominant rhombohedron is chosen as {10$\bar{1}$1}, the other rhombohedron present is {02$\bar{2}$1}, whilst the bipyramid is {22$\bar{4}$3}. There are no faces of a general form present on this crystal.

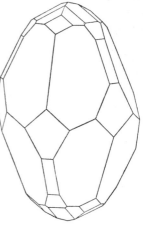

A crystal of calcite of more complex

FIG. 160. A crystal of corundum.

FIG. 161. A crystal of calcite of scalenohedral habit.

development than the simple habit of Fig. 150 is portrayed in Fig. 161. The forms present are the prism {10$\bar{1}$0}, one rhombohedron {$h\,0\,\bar{h}\,l$}, three rhombohedra {$0\,k\,\bar{k}\,l$} and two scalenohedra. The predominance of one of the latter confers a scalenohedral habit on the crystal.

AN ALTERNATIVE METHOD OF INDEXING TRIGONAL CRYSTALS

Miller himself did not use a four-index notation in the hexagonal and trigonal systems, but used a three-index notation throughout. In relation to crystals with a true hexad axis his procedure was very clumsy, and it is now never used in the hexagonal system. In the trigonal system, however, his method has certain advantages over the Miller-Bravais method in some kinds of crystallographic problem and we shall explain it briefly here. His choice of crystallographic axes represents a departure from the recommendation embodied in our formulation of the Law of Rational Indices (p. 40), that ' it is convenient where possible to choose these parallel to prominent axes of symmetry ', for if we follow this suggestion it is impossible to select a set of three non-coplanar axes which are symmetrically related to the triad axes. Miller therefore chose crystallographic axes parallel to the three polar edges of the *fundamental rhombohedron* (Fig. 162) (the one which we have indexed in Miller-Bravais notation $\{10\bar{1}1\}$), and thus not parallel to symmetry axes.

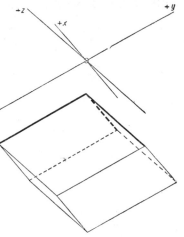

The axes are equally inclined to the triad axis and are non-orthogonal, but make equal angles with each other; this angle between the axes is the plane angle of the face of the fundamental rhombohedron (not the crystallographic interfacial angle), and it depends upon the shape of that rhombohedron in the particular substance in question. Instead of a characteristic axial ratio for each substance, we therefore have in this method of description a characteristic *axial angle* α.

FIG. 162. Miller axes of a trigonal crystal, parallel to the edges of the fundamental rhombohedron.

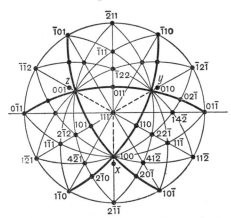

FIG. 163. The upper hemisphere of the projection in Fig. 151 indexed in Miller notation.

The crystals are still set up as before with the triad axis vertical, and are projected on a plane normal to the triad axis. Since the edges of the fundamental rhombohedron define the directions of the crystallographic axes, the indices of the three upper faces must be 100, 010 and 001 (Fig. 163). Notice, however, that the three axes do not emerge through the poles of these faces, since they are parallel to edges and not to face normals. The points of emergence of the axes $x\,y\,z$ can be located in the projection by finding the poles of the zones 010–001,

001–100 and 100–010 respectively. Bearing in mind that the three axes are equally inclined to the plane of projection, we can easily determine the Miller indices of some of the forms which we have already described in the Miller-Bravais notation. Thus:

$$\{0001\} \equiv \{111\},$$
$$\{11\bar{2}0\} \equiv \{10\bar{1}\},$$
$$\{10\bar{1}0\} \equiv \{2\bar{1}\bar{1}\}.$$

The indices of further forms can be obtained by the process of adding indices in two zones (p. 53) and one example will suffice. The pole of a face of the rhombohedron complementary to {100} lies in a zone with 111 and 11$\bar{2}$, and also in a zone with 100 and $\bar{1}2\bar{1}$ (Fig. 163). Adding the first pair gives 22$\bar{1}$, and since this is also in a zone with the second pair $(300 + \bar{1}2\bar{1} = 22\bar{1})$ it is the required index. The two complementary rhombohedra are thus indexed as {100} and {22$\bar{1}$} respectively, and this distinction is often a great advantage when studying trigonal crystals, in which these two forms are quite differently related to the underlying structure. (On the other hand, it was a grave disadvantage in the hexagonal system, where the adjacent faces of a single form, a hexagonal bipyramid, acquired two such different-looking indices as 100 and 22$\bar{1}$; as mentioned above, this notation is now never used for truly hexagonal crystals.)

It may sometimes be necessary to convert an index $p\,q\,r$ of a face in Miller notation to the corresponding index $h\,k\,i\,l$ in Miller-Bravais notation, or *vice versa*. This is readily accomplished on a stereogram, and the faces of a number of forms on the upper hemisphere have been indexed in Fig. 163. If we adopt the convention that the particular face 100 in the one notation shall always be indexed 10$\bar{1}$1 in the other, the following conversions may be useful:

$$h = p - q \qquad k = q - r \qquad i = r - p \qquad l = p + q + r.$$
$$p = h - i + l \qquad q = k - h + l \qquad r = i - k + l.$$

GNOMONIC PROJECTION

We have used stereographic projection exclusively to illustrate the foregoing discussions, and its special advantages should by now be more apparent to the reader. In studies of zonal relationships and problems related to the allocation of indices an equally clear picture is presented by a gnomonogram, and we add here a few examples of this type of projection. Fig. 164 is a gnomonic projection of the crystal of tetragonal mercurous chloride illustrated in Fig. 115. The gnomono-

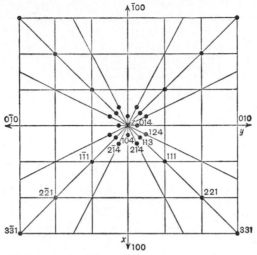

FIG. 164. Gnomonogram of the crystal of mercurous chloride in Fig. 115.

gram of a trigonal crystal (Fig. 165) should be compared with the corresponding stereographic representation (Fig. 151). In a gnomonic projection of a crystal on a plane normal to a prominent axis of symmetry it is usually convenient to complete only a segment of the

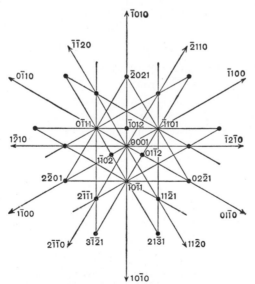

FIG. 165. Gnomonogram of a holosymmetric trigonal crystal.

projection; Fig. 166 is a quadrant of a gnomonogram of the complex orthorhombic sulphur crystal (Fig. 124). To determine the index of

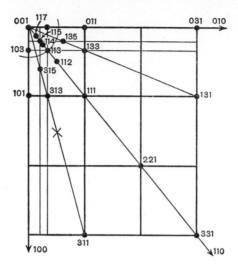

FIG. 166. Gnomonogram of the crystal of sulphur in Fig. 124.

a face projecting in a position such as the one marked by a cross in Fig. 166, we write down the x and y coordinates in terms of the unit distances 001–101 and 001–011 and add 1 as the z index, clearing of fractions if necessary. For the position marked, the index is $1\frac{1}{2}$ $\frac{1}{2}$ 1, i.e. 312.

In order to indicate the scale of a gnomonic projection, a *fundamental circle* (or *ground circle*) of radius r (p. 30) is conventionally added, as in Fig. 166. This circle, of course, is not the gnomonic projection of a great circle on the sphere of projection but of a small circle of ' latitude ' 45°; poles at an angular distance of 45° from P' would lie on it, but in most gnomonograms it has no crystallographic significance.

CHAPTER V

GONIOMETRY

PROCEDURE IN OPTICAL GONIOMETRY

The student should by now be occupied with measurement and projection of actual crystals, and we insert here a chapter describing the procedure to be followed in optical goniometry.

From the crop of crystals under investigation we select a few specimens which appear (to the naked eye or under a lens) to show well-developed plane faces. The number of individual crystals to be measured in a given instance must vary with the nature of the problem in hand, but the earlier tendency to measure ten or even twelve crystals from every crop has now given way to a realisation that the objects of crystal measurement are usually satisfactorily achieved in a much shorter time by the measurement of only two or three carefully selected crystals.

Freehand sketches must be made of each crystal before beginning measurement—plans and elevations from different aspects will serve—and a letter (or number) assigned to every identifiable face so that subsequent readings on the goniometer can be allocated to the correct face and the measurements of successive zones be correctly correlated with each other. If the crystal shows one well-developed zone this should be chosen tentatively as a prism zone and set vertically. The sketch completed, measurement is conducted by investigating one zone at a time, beginning with the most prominent.

To adjust a zone, the crystal is mounted on the goniometer head by means of a small pellet of wax or plasticine, the zone-axis in question being set as nearly as possible parallel to the axis of the instrument and a prominent face in the zone orientated so that its plane is as nearly as possible parallel to the plane of movement of one of the adjusting screws (p. 17). The crystal is centred approximately on the axis, and the image from this first face located. The face is then set accurately normal to the graduated circle by bringing the image centrally on the origin (the fixed slit or cross-wires according to the type of instrument), using for this setting *the other adjusting screw*. A second non-parallel face in the zone, preferably at an angle of between 60° and 90° to the first, is then adjusted in similar manner, but *using only the first adjusting screw* to which the first face was set parallel. In this way the setting of the first face suffers little or no disturbance, though we must turn back to its image and correct the setting if necessary by a touch on the second adjusting screw. By this means, using only the adjustment appropriate to the particular face, the two faces are soon set accurately normal to the graduated circle, and the whole zone is therefore ready for measurement, with its axis parallel to the axis of rotation. If the crystal has been noticeably displaced from the actual axis during the setting, it is re-centred before measurement is begun.

Measurement involves merely bringing the image of each face in turn accurately on the origin, by rotation of the graduated circle, and entering the corresponding readings against a tabulation of the appropriate letters assigned to the zone in the sketch. (As a practical point, it is wise to acquire the habit of rotating the circle in such a direction that these readings decrease as they are recorded downwards on the paper.) The measurement is completed by a second reading from the first face measured, as a check that the crystal has not been displaced and to ensure that every face in the zone has been afforded an opportunity to reflect. During the measurement, the reading only is recorded opposite the corresponding face-letter; subtraction of these readings to give interfacial angles is carried out later. A brief experience of goniometry, however, will soon reveal that the images afforded by crystal faces in practice are of very varied quality, and this quality must be recorded for each reading either in words or by any convenient device such as one or more underlinings for readings from particularly sharp images and one or more sets of brackets around those corresponding to diffuse or otherwise unsatisfactory images.

Further zones are adjusted and measured in a similar manner, until every visible face has been included in at least one zone. The number of zonal measurements necessary will vary with the complexity of the crystal habit; at least three suitably related zones are obviously necessary before the crystal can be projected without unjustified assumptions. (The student must beware, for example, of assuming that an apparent ' base ' on a prismatic crystal is really normal to the prism zone without showing by measurement that it makes an angle of 90° with two non-parallel faces in that zone.) If several crystals are to be measured, each is treated similarly in turn, and averaged angles may be used in the construction of the stereogram. It is in the selection of angles to be used in striking an average that the notes on the quality of images first become useful, and more reliance should be placed on one difference obtained by subtracting two readings corresponding to excellent images than on a number of differences derived from doubtful images.

Images for which satisfactory readings are difficult to obtain may be of three kinds. *Coloured images* should always be ignored; they may arise by total internal reflection after refraction through the crystal or by diffraction from a very narrow or striated face, and from our present point of view are of no significance. *Blurred and distorted images* arise from departures of a particular face from a true plane, due either to imperfections of crystal growth or to subsequent attack by a solvent; so far as possible, crystals with such defective faces should not be used for goniometrical study. *Multiple images*, consisting for example of three or four sharply-defined images regularly grouped in a triangle or square, indicate that an apparent plane face on the crystal actually consists of several portions not quite coplanar. Such are called *vicinal planes* and are of great interest in the study of crystal growth. They sometimes depart from a coplanar disposition only by a fraction of a degree, and if indices were assigned to them these would be fantastically high. We may look upon them at present as an indication of slight departures from perfectly regular repetition of the unit pattern. When they are encountered during measurement the group should be set symmetrically on the cross-wire, and a note or sketch to describe its appearance added to the reading.

From the disposition of the poles on the stereogram the crystal symmetry and hence the system are determined. There follows the most convenient choice of axes, and only then is a parametral plane chosen and the indices of the remaining faces determined. This determination and that of the *axial constants* (the appropriate axial angles and axial ratio or ratios) we can at present only effect graphically. The greater accuracy of optical goniometry over contact goniometry justifies the use of more precise methods, and we shall discuss later the appropriate methods of calculation.

AN EXAMPLE OF A SET OF MEASUREMENTS

The crystal (of barium sulphate) shows an elongated habit, flattened parallel to one pair of faces so as to be almost tabular. We select the long edges tentatively as prism edges, set them vertically and draw the crystal in this attitude. The prism zone has only six faces, so that one elevation together with a plan will serve to show all the faces on the crystal (Fig. 167). We next letter the faces, avoiding pairs of letters (such as *e* and *l* if small letters are used) likely to cause confusion later. The oblique fracture on the lower end of the crystal seems to be largely

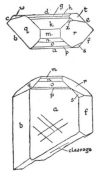

determined by a series of cleavage planes, which appear also as cracks in the body of the crystal. A note is made of this in the drawing, for we shall probably obtain reflections from these planes in the course of measurement and thus acquire further crystallographic information. The shape of this particular crystal will enable us easily to correlate its attitude on the goniometer at any time with the attitude shown in the drawings, but with a very symmetrically-developed crystal we must note carefully in the drawings any visible markings or flaws. In extreme cases it may be necessary to spot one face with Indian ink to avoid any risk of losing the correct orientation.

FIG. 167. Sketched plan and elevation of a crystal to be measured.

Measurement begins with the prism zone *abcdef*. In the actual investigation on which this description is based the large face *a* was set parallel to the plane of movement of one of the adjusting screws, but the image afforded by it was diffuse, and so the crystal was twisted on the wax and the faces *b* and *c* (at an angle approaching 90°) were used for adjustment of the zone.

$$
\begin{array}{ll}
a & (71°\ 8') \\
 & \qquad 52°\ 45' \\
b & 18°\ 23' \\
 & \qquad 74°\ 32' \\
c & \underline{303°\ 51'} \\
 & \qquad 52°\ 41' \\
d & 251°\ 10' \qquad\qquad \dots\dots\dots\dots\dots(1) \\
 & \qquad 52°\ 46' \\
e & 198°\ 24' \\
 & \qquad 74°\ 35' \\
f & \underline{123°\ 49'} \\
 & \qquad 52°\ 42' \\
a & (71°\ 7')
\end{array}
$$

Note that the measurements return to the face from which they start, a second reading being taken from the face *a*. The image afforded by this face is marked as the least satisfactory in the zone whilst those from the pair of parallel faces *c* and *f* were exceptionally sharp and clear.

Subtracting the readings to give the interfacial angles we naturally look with interest for any evidence of symmetry, but it cannot be too strongly emphasised that the proper stage for the deduction of symmetry is not reached until one has constructed a projection showing *all* the faces present on the crystal. A prism zone, for example, showing a succession of faces at angles within a few minutes of 45° may have nothing to do with cubic or tetragonal symmetry, a conclusion to which the student is apt to jump at this stage, but may be revealed later as merely a pseudo-tetragonal zone in, perhaps, a monoclinic crystal. Before passing to the next zone we may add up the interfacial angles to check that the sum differs from 360° only by the difference between the two readings for the face *a*; this is a check only on our arithmetic, but experience proves it to be a useful one!

The next prominent zone is *dghkmnopa* with its obvious parallelism of edges. The faces *d* and *m* (avoiding *a*) may be used for adjustment.

$$
\begin{array}{lll}
d & 87°\ 47' & \\
 & & 21°\ 56' \\
g & 65°\ 51' & \\
 & & 16°\ 54' \\
h & 48°\ 57' & \\
 & & 19°\ 19' \\
k & 29°\ 38' & \\
 & & 31°\ 50' \\
m & (357°\ 48') & \\
 & & 31°\ 48' \qquad \ldots\ldots\ldots\ldots\ldots(2) \\
n & 326°\ \ 0' & \\
 & & 19°\ 19' \\
o & 306°\ 41' & \\
 & & 16°\ 55' \\
p & 289°\ 46' & \\
 & & 21°\ 56' \\
a & (267°\ 50') & \\
 & & 180°\ \ 2' \\
d & 87°\ 48' &
\end{array}
$$

With the exception of *m*, all the faces gave satisfactory images; if *m* had been of still poorer quality it might have been necessary to avoid using it for adjusting the zone, but there is no difficulty in choosing a suitable pair of faces in such a highly modified zone.

The choice of the next zone for measurement deserves some consideration. Notice that measurement of *qmr* would not be particularly useful. The zone would clearly not pass through any prism face, so that it would not serve to determine the relationship of the two zones already measured; nor would it suffice to determine the positions of *q* and *r* without further measurement. We therefore decide to measure the angular distance of *r* from the good prism face *f*, expecting from our knowledge of the prevalence of simple zonal relationships that the zone *fr* will pass through one of the faces *k*, *h* or *g* (probably through *k* to judge by the apparent parallelism of edges visible on the crystal), and possibly also through *w*. Adjusting the zone by means of *f* and *r*, we find that *k* does lie in this zone, and if necessary we can use *k* and *f* for the final adjustment.

$$
\begin{array}{lll}
c & 237° 54' & \\
 & & 71° 12' \\
k & 166° 42' & \\
 & & 49° 2' \\
r & 117° 40' & \\
 & & 59° 48' \\
f & 57° 52' & \\
 & & 179° 58' \\
c & 237° 54' &
\end{array}
\qquad \ldots\ldots\ldots\ldots\ldots\ldots(3)
$$

Evidently the face *w* does not fall in this zone, since no reflection was seen in passing across that part of the crystal.

The symmetry suggested by the appearance of the crystal in plan prompts us next to set the zone *cq*, and this proves to pass through *n*.

$$
\begin{array}{lll}
f & 70° 11' & \\
 & & 71° 25' \\
n & 358° 46' & \\
 & & 48° 51' \\
q & 309° 55' & \\
 & & 59° 47' \\
c & 250° 8' & \\
 & & 120° 8' \\
\text{cleavage } ((130° \ 0')) & \\
 & & 59° 50' \\
f & 70° 10' &
\end{array}
\qquad \ldots\ldots\ldots\ldots\ldots\ldots(4)
$$

The cleavage surfaces do not give a very sharp reflection, so that the reading is somewhat uncertain, but it clearly corresponds to a cleavage parallel to the face *q*.

The faces *w*, *t*, *s* and *x* are not yet included in any of the measured zones. We follow once more the same kind of argument. To fix *s*, for example, it may be necessary to measure in turn the zones *fsc* and *asd* and so locate *s* at the intersection of two small circles of radii *fs* and *as* about *f* and *a* respectively. It is extremely probable, however, that the zone *fsc* will prove to include some other face or faces already measured; setting the zone on the goniometer, we find it passes also through *o*.

$$
\left[
\begin{array}{ll}
f & 310°\ 36' \\
 & \qquad 26°\ 36' \\
s & 284°\ \ 0' \\
 & \qquad 35°\ 13' \\
o & 248°\ 47' \\
 & \qquad 118°\ 14' \\
c & 130°\ 33' \\
 & \qquad 179°\ 59' \\
f & 310°\ 34'
\end{array}
\right] \quad \dots\dots\dots\dots(5)
$$

We next fix the face *t* in a similar manner.

$$
\left[
\begin{array}{ll}
b & 254°\ 22' \\
 & \qquad 118°\ 14' \\
h & 136°\ \ 8' \\
 & \qquad 35°\ 10' \\
t & 100°\ 58' \\
 & \qquad 26°\ 37' \\
e & \ 74°\ 21' \\
 & \qquad 180°\ \ 0' \\
b & 254°\ 21'
\end{array}
\right] \quad \dots\dots\dots\dots(6)
$$

Since *w* appears to be similarly situated on the other side of the crystal we next set the zone *cw*.

$$
\left[
\begin{array}{ll}
c & 264°\ \ 8' \\
 & \qquad 26°\ 40' \\
w & (237°\ 28') \\
 & \qquad 35°\ 11' \\
h & 202°\ 17' \\
 & \qquad 39°\ 30' \\
x & (162°\ 47') \\
 & \qquad 78°\ 41' \\
f & \ 84°\ \ 6' \\
 & \qquad 179°\ 58' \\
c & 264°\ \ 8'
\end{array}
\right] \quad \dots\dots\dots\dots(7)
$$

This proves to pass through the face *x*, so that all the faces have now been located.

To make the projection, we first plot zone (1) around the primitive, retaining an orientation to correspond with our original drawings. From zones (2) and (3) we know that $dk = 58°\ 9'$ and $ck = 71°\ 12'$, and *k* is thus located at the intersection of small circles of these radii about *d* and *c* respectively. (A small circle of radius more than 70° has a rather large radius in projection, so that it is convenient in practice to draw only the circle of radius 58° 9′ and to prick through the point of intersection from a superposed stereographic net.) The face k having been inserted, we can draw in turn the zones (2) (which proves to be vertical) and (3) and plot the faces upon them. Finally the zones (4), (5), (6) and (7) can be plotted. The stereogram completed to this stage is shown in Fig. 168.

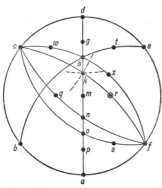

FIG. 168. Stereogram constructed from measured zones.

The high quality of most of the reflections and the excellent agreement of the readings leave us in no doubt, in this particular instance, about the positions of any of the faces. Suppose, however, that the face x had been of much poorer quality and only showed a brightening without a definite image during the measurement of the zone $cwhxf$. This shows that it lies in that zone, somewhere between h and f, but we might be in doubt about its exact position. The projection may then afford suggestions for further zonal measurements. We see in Fig. 168 that x appears to lie in the zone $exmb$ and also in the zone $dxra$ (this can be checked by superposing a net) and either or both of these zones could be set up and measured; since both m and a give rather poor reflections there is little to choose between them.

Having completed the projection we are in a position to determine the symmetry. The symmetrical distribution of the pairs of faces k and n, h and o, g and p about m indicates the presence of a diad axis normal to m. The stereogram shows also bilateral symmetry about the zone dma; there must be a vertical plane of symmetry passing through the diad axis, and this combination automatically gives rise to a second plane of symmetry at right angles to the other (and hence coinciding with the zone circle qmr). Since the lower portion of the crystal is terminated only by cleavages we cannot determine by goniometry whether there is a centre of symmetry, though the determination is easily made by other methods (see p. 155). If we assume for the present that the crystal is centro-symmetrical, there will arise also a horizontal plane of symmetry at right angles to the vertical diad axis and also two horizontal diad axes normal to the vertical planes of symmetry— on this assumption, the crystal is orthorhombic holosymmetric.

The three diad axes will be chosen as the directions of crystallographic x, y and z axes. On the evidence of this one crystal there is no reason for not choosing the vertical diad axis as the z direction (in the conventional orientation of barium sulphate it has been taken as the x direction). Of the horizontal diad axes either can be chosen as the

x direction, but to enable us to index e as 110 and still determine an axial ratio a/b less than unity (p. 71) it is perhaps preferable to choose the normal to d as the y axis. The projection is reproduced in Fig. 169 in this orientation. The face d, parallel to the x and to the z axes, is 010 and m is 001.

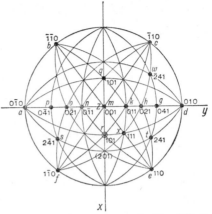

FIG. 169. The stereogram in Fig. 168 reorientated after choice of x, y, z directions. Extra zone circles have been drawn for the graphical determination of indices.

We next choose a parametral plane; x and t are the only possibilities, of which we select x as giving the simpler indices to the remaining forms, for the zonal relationships show at once that e is then 110, k is 011 and r is 101. (We have admittedly thus chosen as parametral plane a face which is morphologically very unimportant, but this can be justified by the prominent development of the forms {101} and {011} following this choice.) The face h, lying in a zone between 011 and 010 and also in a zone between 111 and $\bar{1}$10, must be 021. The face t lies in a zone between 021 and 110, but we must guard against making the mistake of indexing this as 131, for 131 would lie in a zone between 010 and 111. Actually t lies in a zone between 010 and 201 (not present on the crystal, but indexed in brackets in Fig. 169); it is therefore the face 241. Finally, g, in a zone between 021 and 010 and also between 241 and $\bar{2}$41, is the face 041.

All the faces present have now been indexed by graphical determination, though after we have studied crystallographic calculations it would often be quicker to determine some of these indices more directly from the measured angles. Determination of the axial ratios, also, would often need a knowledge of such calculations; in this particular instance we can determine them directly from measured angles and thus complete our description of the crystal, though the student may prefer to omit the remainder of this chapter until he has studied Chapter VIII.

The axial ratio a/b is determined by the slope of the face 110 in relationship to the x and y axes:

$$a/b = \tan 100 \frown 110 = \tan \tfrac{1}{2}(\text{average values of } bc \text{ and } ef)$$
$$= \tan 37° \ 17'$$
$$= 0\cdot761$$

Similarly, the ratio c/b is determined by the slope of the face 011 in relationship to the z and y axes:

$$c/b = \tan 001 \frown 011 = \tan (\text{average value of } mk \text{ and } mn)$$
$$= \tan 31° \ 49'$$
$$= 0\cdot620.$$

DESCRIPTION OF THE CRYSTAL OF BARIUM SULPHATE

Symmetry.   Orthorhombic holosymmetric (assuming the presence of a centre of symmetry).

Axial ratios. $a : b : c = 0\cdot761 : 1 : 0\cdot620$.

Forms represented.   Pinacoids {010}, {001}.
Prism {110}.
Domes {011}, {021}, {041}, {101}.
Bipyramids {111}, {241}.

The form {101} is a cleavage form.

(This description differs from the one given in standard texts, owing to the different choice of axes mentioned above.)

CHAPTER VI

THE THIRTY-TWO CLASSES

THE THIRTY-TWO CLASSES OF CRYSTAL SYMMETRY

In our first approach to the study of crystal symmetry we utilised only the simple symmetry elements—a centre, one or more planes of reflection symmetry, and one or more rotary axes of degree 2, 3, 4 or 6. At the same time it was stated that there were in all thirty-two different crystal classes (possible combinations of these symmetry elements, together with a completely asymmetric class). Actually it is possible to build up only thirty such classes on the basis of these simple assumptions, and for this and other reasons we must modify this description slightly before proceeding further. The centre of symmetry, as a fundamental element, is now no longer used, though the concept of centrosymmetry is still useful; crystal symmetry is described in terms of:

one or more planes of reflection symmetry, symbolised m;
rotation axes, symbolised in terms of their degree 1, 2, 3, 4 or 6;
axes of rotary inversion (or *inversion axes*), symbolised also in terms of their degree, $\bar{1}$, $\bar{2}$, $\bar{3}$, $\bar{4}$ or $\bar{6}$.

The new symmetry elements, inversion axes, are *compound symmetry elements*. They carry out on a given crystal plane the operation of rotation through the angle indicated by the degree together with inversion across a centre. Thus an axis $\bar{4}$ (read 'bar four') normal to the paper operates on a pole 1 (Fig. 170) to rotate it through 90° (to a position above the ring 4), followed by inversion to the position 2; this compound operation is then repeated until the original position is again reached. Thus from position 2 the pole is rotated a further 90° and inverted to position 3; rotated a further 90° and inverted to position

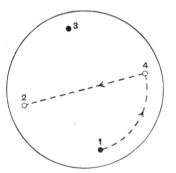

FIG. 170. Stereogram to show the operation of an inversion tetrad axis $\bar{4}$.

4; rotated a further 90° and inverted to resume position 1. The final disposition of two planes above the paper and two symmetrically related planes below is a crystallographically possible arrangement which has more than twofold symmetry and yet clearly does not possess a true tetrad axis. The symmetry, in fact, is that of one of the thirty-two classes which we could not have derived from our simple approach to the study of crystal symmetry.

Fig. 171 shows by the distribution of poles on stereograms the operation of the remaining inversion axes, and it will be seen that

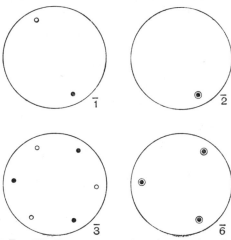

FIG. 171. Stereograms to show the operation of inversion axes $\bar{1}, \bar{2}, \bar{3}$ and $\bar{6}$.

each of these could also be described in terms of the simple elements which we first utilised. Thus $\bar{1}$ produces the same distribution of planes as does a centre of symmetry. $\bar{2}$ is equivalent to a reflection plane in the plane of projection (i.e. $\bar{2} \equiv m$). $\bar{3}$ is equivalent to a triad axis combined with a centre. $\bar{6}$ is equivalent to a triad axis normal to a plane of symmetry. Having agreed to abandon the use of the centre as an independent element of symmetry in our descriptions, we shall in future symbolise all these symmetry groups in terms of the appropriate inversion axis. It should be clearly understood that this particular choice is made only for convenience, and that there are often many *possible* alternative descriptions of a symmetry of distribution. In the past, indeed, another kind of compound symmetry element, an *alternating axis*, was used in place of inversion axes. Such an axis combined a rotation with a reflexion across a plane normal to the axis. It will be a useful exercise for the student to convince himself of the following identities:

an alternating axis $1 \equiv \bar{2}$,
an alternating axis $2 \equiv \bar{1}$,
an alternating axis $3 \equiv \bar{6}$,
an alternating axis $4 \equiv \bar{4}$,
an alternating axis $6 \equiv \bar{3}$,

but the concept of alternating axes should be abandoned in favour of that of inversion axes in modern crystallography.

On this basis we proceed to build up the thirty-two crystal classes systematically. The principal axis (normal to the paper in a projection) is first set down; it may be either a rotation axis or an inversion axis and is denoted by the appropriate symbol. If there is a reflection plane normal to this axis, the symbol m is added as $\frac{2}{m}$ (read ' two over em ', but usually printed for convenience $2/m$). A reflection plane through the axis is written without the stroke ($2m$, read ' two em '). If both kinds of plane are present the symbol is $2/mm$. A horizontal diad axis (normal to the principal axis) is indicated by adding a figure 2 (as 32, which must be read as ' three two ' and not as thirty-two).

Using the general symbol X to denote a principal axis of any degree we may have the following combinations:

X rotation axis alone.
$\bar{X}$ inversion axis alone.
X/m rotation axis normal to a plane of symmetry.
Xm rotation axis with a vertical plane of symmetry.
$\bar{X}m$ inversion axis with a vertical plane of symmetry.
$X2$ rotation axis with a diad axis normal to it.
X/mm rotation axis with both kinds of plane of symmetry.

(It may be observed that the symbols $\bar{X}/m$ and $\bar{X}2$ do not appear; why not?)

The chart on p. 106 shows the complete scheme of thirty-two classes, illustrated by stereograms of the general forms, developed on this principle. Beneath each stereogram is the appropriate symbol for the class, and it will be seen that this symbol is not in every instance precisely the one which would be derived from the position of the class on the chart. Where the two differ considerably, the systematic symbol is placed in brackets on the right whilst the customary symbol is un-bracketed. The class $2m$, for example, is to be symbolised mm, and the class $2/mm$ is always denoted mmm. These changes are made in view of the demands which we shall presently make on this notation when we come to study the vastly greater number of possible types of arrangement in the *internal* structure of crystals. The changes consist mainly of the use, in the customary symbol, of elements of symmetry which arise automatically from combination of those indicated by the systematic symbol. The class $2/mm$, for example, can be seen from its stereogram to be the orthorhombic bipyramidal class which we have already studied, and the customary symbol mmm lists the three planes

THE 32 CRYSTAL CLASSES

of symmetry which we know this class to possess. Other similar changes will be clear when we have studied the classes in detail.

The grouping of the classes into systems also requires some explanation, for it is apparently not quite in accord with our earlier statements. Classes 1 and $\bar{1}$ in the triclinic system call for no comment. Classes 2, m and $2/m$ in the monoclinic system are correctly described as possessing ' one diad axis ' (p. 7) now that we recognise m as the equivalent of an inverse diad axis. Of the classes mm, 222 and mmm in the orthorhombic system, mm shows only one true diad axis, though we may consider the two planes of symmetry as equivalent to $\bar{2}$ axes; the class is placed in the orthorhombic system because the symmetry of its optical and many other physical properties is the same as that of the two other classes in this system. In the trigonal system are five classes, each of which possesses an axis 3 or $\bar{3}$; the classes $3/m$ and $3/mm$ are assigned to the hexagonal system, since the principal axis in each has the symmetry $\bar{6}$. The seven classes in each of the tetragonal and hexagonal systems then call for no further comment. The cubic system contains five classes of which each symbol includes a figure 3, denoting a *secondary* triad axis (a triad not in the position of the principal axis, since it does not appear first in the symbol). Further details of the application of this notation to the classes of the cubic system will be given a little later (p. 141) when these are described in detail.

We shall now discuss each class of symmetry in turn, describing the possible forms and mentioning some substances which crystallise in the class in question. It is convenient to reverse the procedure of Chapter IV and, following the chart, to build up the symmetry gradually, beginning with the completely asymmetric class.

TRICLINIC SYSTEM

CLASS 1. (Asymmetric, triclinic pedial.) No symmetry. The crystallographic axes are chosen parallel to any suitable edges (Fig. 172).

Forms. There are no special forms. Every form consists of a single face, a *pedion* ($\pi\epsilon\delta\acute{\iota}o\nu$, a plain), and an actual crystal must show at least four forms.

Examples. The substance usually quoted as an example of triclinic pedial symmetry is calcium thiosulphate, $CaS_2O_3 . 6H_2O$. Two described habits are illustrated in Figs. 173, 174, but we shall see later that it is not easy always to be quite certain of the correct assignment of a crystal to this class, and the examples figured may be distorted habits of crystals showing pinacoidal symmetry (class $\bar{1}$ below).

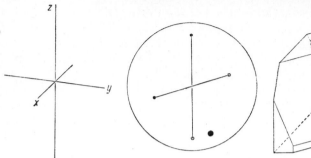

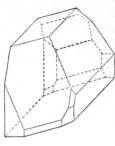

FIG. 172. Class 1; the crystallographic axes and a stereogram of the general form. (See the note to the legend of Fig. 127.)

FIG. 173. A crystal of calcium thiosulphate.

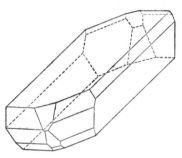

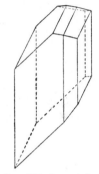

FIG. 174. Another habit of calcium thiosulphate.

FIG. 175. A crystal of rubidium ferrocyanide.

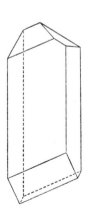

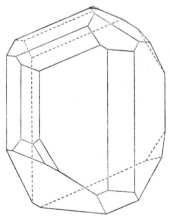

FIG. 176. A crystal of strontium hydrogen tartrate.

FIG. 177. Another habit of strontium hydrogen tartrate.

Other examples are rubidium ferrocyanide, $Rb_4Fe(CN)_6 . 2H_2O$ (Fig. 175); strontium tartrate, $SrH_2(C_4H_4O_6)_2 . 4H_2O$ (Figs. 176, 177). Amongst minerals, parahilgardite, $Ca_8B_{18}O_{33}Cl_4 . 4H_2O$, is placed here.

CLASS $\bar{1}$. (Triclinic holosymmetric, triclinic pinacoidal.) An inversion identity axis equivalent to a centre of symmetry (Fig. 178);

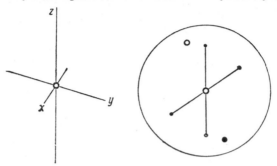

Fig. 178. Class $\bar{1}$; the crystallographic axes and a stereogram of the general form. The centre of symmetry is denoted by a thick ring.

as in class 1, the crystallographic axes are chosen parallel to any suitable crystal edges.

Forms. There are no special forms. Every form is a *pinacoid*.

Examples. Copper sulphate, $CuSO_4 . 5H_2O$; potassium persulphate, $K_2S_2O_8$ (Fig. 179); bismuth nitrate, $Bi(NO_3)_3 . 9H_2O$ (Fig. 180).

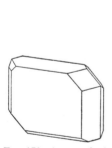

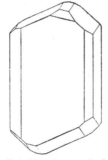

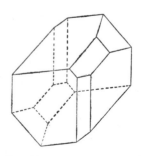

Fig. 179. A crystal of potassium persulphate

Fig. 180. A crystal of bismuth nitrate.

Fig. 181. A crystal of axinite.

Amongst minerals the plagioclase felspars, an isomorphous series between albite $NaAlSi_3O_8$, and anorthite $CaAl_2Si_2O_8$, are all triclinic pinacoidal; kyanite, Al_2SiO_5; sassoline, $B(OH)_3$. The mineral axinite, a complex borosilicate, is often placed here, an idealised crystal being represented as in Fig. 181, but the crystals show the property of pyroelectricity (p. 155) and must be assigned to class 1.

MONOCLINIC SYSTEM

CLASS 2. (Monoclinic hemimorphic, monoclinic sphenoidal.) One diad axis (always chosen as the y crystallographic axis, Fig. 182).

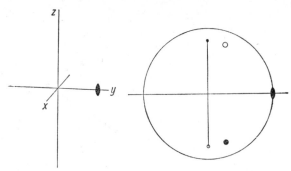

FIG. 182. Class 2; the crystallographic axes and a stereogram of the general form.

Special forms.   Pedions {010}, {0$\bar{1}$0}.
Pinacoids {h 0 l}.
General forms.   Sphenoids {h k l}.

The open wedge-like form obtained by repetition of a face around a diad axis is called a *sphenoid* (σφήν, a wedge). Geometrically the pair of faces resembles those of a dome, but this term is restricted to a form developed by reflection over a plane. Note that the forms {100} and {001} are included amongst the pinacoids {h 0 l}, and the forms {h k 0} amongst the sphenoids {h k l}, since of the crystallographic axes x, y, z only y is a symmetry axis.

Examples. Lithium sulphate, $Li_2SO_4 . H_2O$; sucrose and some other sugars; quercitol, $C_6H_7(OH)_5$; tartaric acid $COOH(CHOH)_2COOH$

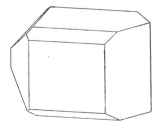

FIG. 183. A crystal of tartaric acid.

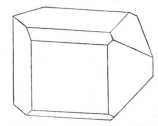

FIG. 184. A crystal of tartaric acid which is the enantiomorph of the crystal in Fig. 183.

(Figs. 183, 184). The figures of this substance illustrate clearly two important features of the crystallography of this class. First, there is a different facial development at opposite ends of the diad axis (the y crystallographic axis, running left and right); such an axis is said to be uniterminal (p. 61) or *polar*, and the class may be called the hemimorphic class of the monoclinic system. Important physical properties which we shall mention later are associated with this type of symmetry. Secondly, the crystal can show either of two different aspects which are *enantiomorphous*; they are mirror images of each other, and like a right and a left hand are not superposable in space. This feature, also, is related to a particular type of symmetry and is associated with special physical properties (p. 153).

Other examples in this class are ethylammonium bromide, $NH_3C_2H_5Br$, and the corresponding iodide, $NH_3C_2H_5I$ (Fig. 185). Very few minerals belong here, the best example being pickeringite, $MgSO_4 . Al_2(SO_4)_3 . 22H_2O$.

FIG. 185. A crystal of ethylammonium iodide.

CLASS m. (Monoclinic clinohedral, monoclinic domatic.) An inversion diad axis (equivalent to a plane of symmetry, which is always set normal to the y crystallographic axis, Fig. 186).

Special forms.   Pedions $\{h\ 0\ l\}$.
                    Pinacoid $\{010\}$.
General forms. Domes $\{h\ k\ l\}$.

By analogy with the special forms in the orthorhombic bipyramidal class, which we have already described, the general forms obtained here

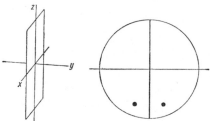

FIG. 186. Class m; the crystallographic axes, showing their relationship to the plane of symmetry, and a stereogram of the general form.

by reflection of a generally-situated face across the plane of symmetry are called domes; they are two-faced open forms, and include $\{h\ k\ 0\}$ and $\{\bar{h}\ k\ 0\}$. The forms $\{100\}$, $\{001\}$, $\{\bar{1}00\}$ and $\{00\bar{1}\}$ are, of course, included amongst the pedions.

Examples. Potassium tetrathionate, $K_2S_4O_6$ (Fig. 187); the sodium silicate, $Na_2SiO_3 . 5H_2O$ (Fig. 188); potassium nitrite, KNO_2. Mineralogical examples are provided by clinohedrite, $H_2CaZnSiO_5$; scolecite,

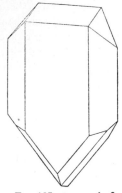

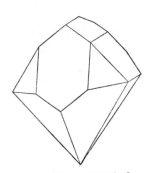

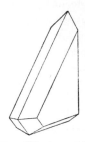

FIG. 187. A crystal of potassium tetrathionate.

FIG. 188. A crystal of $Na_2SiO_3 . 5H_2O$.

FIG. 189. A crystal of hilgardite.

$CaAl_2Si_3O_{10} . 3H_2O$; hilgardite, $Ca_8B_{18}O_{33}Cl_4 . 4H_2O$ (Fig. 189); and the clay minerals kaolinite, nacrite and dickite $Al_2Si_2O_5(OH)_4$.

CLASS $2/m$. (Monoclinic holosymmetric, monoclinic prismatic.) A diad axis (chosen as the y crystallographic axis) normal to a plane of symmetry, involving a centre of symmetry (Fig. 190).

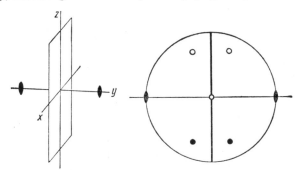

FIG. 190. Class $2/m$; the elements of symmetry, crystallographic axes and a stereogram of the general form.

Special forms.   Pinacoid {010}.
Pinacoids {$h\,0\,l$}.
General forms.   Prisms {$h\,k\,l$}.

This is the class of the monoclinic system which we have already studied, and the student may be reminded of the more usual informal nomenclature (p. 76).

Examples.   Very many substances crystallise with the symmetry of this class, and we can quote only a few typical examples: sodium carbonate, $Na_2CO_3 . 10H_2O$; sodium bicarbonate, $NaHCO_3$ (Fig. 132);

potassium chlorate, $KClO_3$; numerous sulphates such as $FeSO_4 . 7H_2O$ and double sulphates and selenates such as $(NH_4)_2Mg(SO_4)_2 . 6H_2O$; a wide range of organic compounds such as naphthalene, anthracene and glycine. Amongst minerals we have already figured borax (Fig. 125), gypsum (Fig. 131) and trona (Fig. 133); here belong also a great number of rock-forming minerals such as the micas, chlorites and many members of the families of pyroxenes, amphiboles and epidotes.

ORTHORHOMBIC SYSTEM

CLASS *mm.* (Orthorhombic hemimorphic, orthorhombic pyramidal.) Two planes of symmetry at right-angles, intersecting in a diad axis (always chosen as the *z* crystallographic axis, Fig. 191).

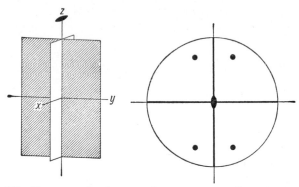

FIG. 191. Class *mm*; the elements of symmetry, crystallographic axes and a stereogram of the general form. The planes of symmetry, parallel to different pinacoids, are differently shaded.

Special forms.   Pedions {001}, {00$\bar{1}$}.
Pinacoids {100}, {010}.
Prisms {*h k* 0}.
Domes {*h* 0 *l*}, {0 *k l*}.

General forms.   Pyramids {*h k l*}.

The diad axis (*z* axis) is here uniterminal, so that the general forms are open four-faced pyramids and {*h k* $\bar{l}$} is a separate form from {*h k l*}.

Examples.   Bismuth thiocyanate, $Bi(CNS)_3$; triphenylmethane, $CH(C_6H_5)_3$; picric acid, $C_6H_2(NO_2)_3 . OH$; resorcinol, $C_6H_4(OH)_2$. As mineralogical examples: pirssonite, $CaNa_2(CO_3)_2 . 5H_2O$ (Fig. 192); struvite, $NH_4MgPO_4 . 6H_2O$ (Fig. 193); hemimorphite (smithsonite, electric calamine), $Zn_4(OH)_2Si_2O_7 . H_2O$ (Fig. 194); bertrandite, $Be_4(OH)_2Si_2O_7$; natrolite, $NaAl_2Si_3O_{10} . 2H_2O$.

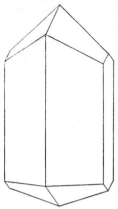

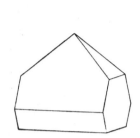

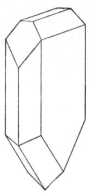

FIG. 192. A crystal of
pirssonite.

FIG. 193. A crystal of
struvite.

FIG. 194. A crystal
of hemimorphite.

CLASS 222. (Orthorhombic sphenoidal.) Three mutually perpendicular diad axes (Fig. 195), always chosen as the directions of the crystallographic axes x, y, z.

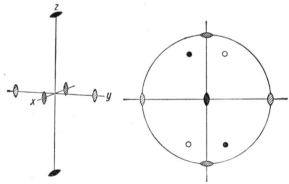

FIG. 195. Class 222; the elements of symmetry, crystallographic axes and a stereogram of the general form. The flags of the three diad axes, normal to different pinacoids, are differently shaded.

Special forms.   Pinacoids {100}, {010}, {001}.
                              Prisms {$h\,k\,0$}, {$0\,k\,l$}, {$h\,0\,l$}.
General forms.   Sphenoids {$h\,k\,l$}.

The three diad axes, which are chosen as the directions of the x, y, z crystallographic axes, are all of similar crystallographic significance, so that there are six possible settings for every crystal in this class. Once a setting has been adopted, it is customary to refer to forms {$0\,k\,l$} and {$h\,0\,l$} as domes, restricting the term prism to forms {$h\,k\,0$} which are developed from faces parallel to the particular diad axis chosen as the *z* direction. The general forms are sphenoids, closed four-faced wedge-like forms resembling the tetrahedron of the cubic system but possess-ing no planes of symmetry (Fig. 196). Some crystallographers call

FIG. 196. Orthorhombic sphenoids {*h k l*} and {*h k̄ l*}.

such a form a *bisphenoid* (or *disphenoid*), regarding it as made up of two of the sphenoidal pairs of faces composing the open sphenoidal forms described in class 2 (the sphenoidal class of the monoclinic system). This usage is inconsistent in view of the meaning which we have attached to these two prefixes; the former implies a plane of symmetry and the latter a repetition of pairs of faces around an axis, and neither of these types of symmetry is displayed by the forms in question. We shall call them simply (*orthorhombic*) *sphenoids*.

Figs. 196, 197 show also that there are two related types of sphenoid in this class, the one right-handed and the other left-handed, so that

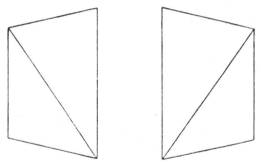

FIG. 197. The sphenoids in Fig. 196 redrawn in a different orientation to display more clearly their enantiomorphous relationship.

the form {*h k l*} is the mirror-image of the form {*h k̄ l*}. This is there-fore a further class in which crystals may exhibit the property of enantiomorphism.

Examples. Typical examples are provided by magnesium sulphate (epsomite), $MgSO_4 . 7H_2O$ (Fig. 198); zinc sulphate (goslarite), $ZnSO_4 . 7H_2O$, and a number of related sulphates and chromates.

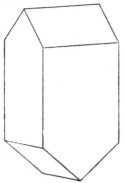

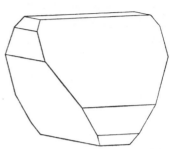

Fig. 198. A crystal of epsomite.   Fig. 199. A crystal of tartar emetic.

Tartar emetic, $K(SbO)C_4H_4O_6$, frequently assumes a markedly sphen-oidal habit (Fig. 199).   The series of tartrates of which Rochelle salt, $KNaC_4H_4O_6 . 4H_2O$, is a member, are usually of prismatic habit, but the presence of sphenoidal forms reveals the true symmetry; in Fig. 200 the forms developed are, on the customary indexing, the pinacoids

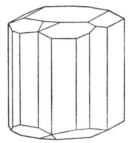

Fig. 200. A crystal of Rochelle salt.   Fig. 201. A crystal of asparagine.

{100}, {010} and {001}; prisms {210}, {110} and {120}; domes {101} and {011}; sphenoids {211} and {1$\bar{1}$1}.   Strontium formate, $Sr(HCOO)_2$ and the similar compounds of barium and lead; methylurea, $CONH_2(NHCH_3)$; asparagine (Fig. 201) and a number of alkaloids such as narcotine, atropine, strychnine and codeine afford further examples.   Amongst minerals, in addition to those already mentioned, may be noted austinite, $CaZn(OH)AsO_4$; olivenite, $Cu_2(OH)AsO_4$; and chalcomenite, $CuSeO_3 . 2H_2O$.

   CLASS *mmm.*   (Orthorhombic holosymmetric, orthorhombic bi-pyramidal.)   Three planes of symmetry intersecting in three mutually perpendicular diad axes (chosen as the directions of crystallographic *x y z*) and a centre of symmetry (Fig. 202).

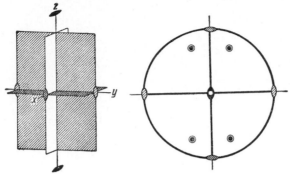

FIG. 202. Class *mmm*; the elements of symmetry, crystallographic axes and a stereogram of the general form.

Special forms. Pinacoids {100}, {010}, {001}.

Prisms {*h k* 0}, {*h* 0 *l*}, {0 *k l*}.

General forms. Bipyramids {*h k l*}.

This is the orthorhombic bipyramidal class, which we have already studied. It may be useful to point out again here that once a particular setting has been adopted the prisms {*h* 0 *l*} and {0 *k l*} are usually called domes.

Examples. This class ranks next in importance to the class $2/m$ from the point of view of the number of substances belonging to it, and we can here only select a few examples at random. Sulphates R_2SO_4 of the alkali metals, and the isomorphous selenates; perchlorates of these metals, and some permanganates; potassium thiocyanate, KCNS, are selections from a wide field of inorganic representatives. Oxalic acid, $(COOH)_2$; thiourea, $CS(NH_2)_2$; many long-chain paraffins such as nonicosane, $C_{29}H_{60}$; *o*-nitraniline, $C_6H_4(NO_2)(NH_2)$, will serve to represent the organic field, whilst as mineralogical examples may be mentioned anglesite, $PbSO_4$ (Fig. 116), and the isomorphous barytes, $BaSO_4$, and celestine, $SrSO_4$; sulphur (Figs. 123, 124); stibnite, Sb_2S_3; brookite, TiO_2; forsterite, Mg_2SiO_4, and other members of the olivine group.

TRIGONAL SYSTEM

CLASS 3. (Trigonal hemimorphic, trigonal pyramidal.) A single triad axis (set vertically, and chosen as the *z* direction in the Miller-Bravais notation, Fig. 203).

		Miller-Bravais	*Miller*
Special forms.	Pedions	{0001}, {000$\bar{1}$}.	{111}, {$\bar{1}\bar{1}\bar{1}$}.
	Trigonal prisms	{*h k i* 0}.	{*p q* −*p*+*q*}.
General forms.	Trigonal pyramids	{*h k i l*}.	{*p q r*}.

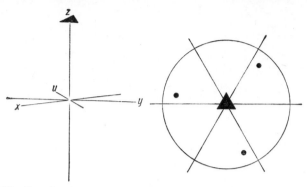

FIG. 203. Class 3; the crystallographic axes, showing their relationship to the triad axis, and a stereogram of the general form.

It will be noticed here (as in many other classes of low symmetry) that the general form considered alone, or in certain combinations, appears to present more symmetry than is proper to the class. Thus a trigonal pyramid closed by a basal pedion appears to show three vertical planes of symmetry, and it is only the combination of two or more general forms, or of a general form with a particular type of special form, which will reveal the true lack of symmetry. In a trigonal prism combined with a trigonal pyramid in this class, the prism edges, for example, need not lie in the same vertical planes as do the edges of the pyramid.

Examples. Sodium periodate, $NaIO_4 . 3H_2O$, was for a long time the only substance placed with certainty in this class. The usual habit displays several trigonal pyramids and a large development of the basal pedion $\{\bar{1}\bar{1}\bar{1}\}$ (Fig. 204); the forms present on the particular example illustrated are: three upper ('positive') pyramids, usually indexed $\{100\}$, $\{110\}$, $\{11\bar{1}\}$; three lower ('negative') pyramids $\{\bar{1}00\}$, $\{1\bar{1}\bar{1}\}$, $\{1\bar{3}\bar{1}\}$; basal pedion $\{\bar{1}\bar{1}\bar{1}\}$; and one trigonal prism $\{10\bar{1}\}$.

FIG. 204. A crystal of sodium periodate.

FIG. 205. A crystal of magnesium sulphite.

Magnesium sulphite, $MgSO_3 . 6H_2O$, and the similar compounds of nickel and cobalt, are now assigned to this class. The crystals are clearly hemimorphic (Fig. 205), but usually lack decisive morphological evidence of the absence of vertical planes of symmetry.

The carbonate $Na_2Mg(CO_3)_2$ may belong here, and a mineralogical example may be provided by the recently-described mineral gratonite, $Pb_9As_4S_{15}$, but a final decision has not proved possible from an examination of the material so far available.

CLASS $\bar{3}$. (Rhombohedral.) An inversion triad axis (equivalent to a rotation triad axis and a centre) (Fig. 206).

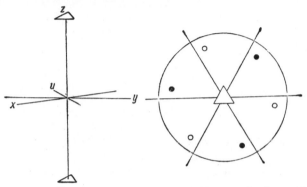

Fig. 206. Class $\bar{3}$; the inversion axis, flagged with open triangles, crystallographic axes and a stereogram of the general form.

		Miller-Bravais	*Miller*
Special forms.	Pinacoid	{0001}.	{111}.
	Hexagonal prisms	$\{h\,k\,i\,0\}$.	$\{p\,q-\overline{p+q}\}$.
General forms.	Rhombohedra	$\{h\,k\,i\,l\}$.	$\{p\,q\,r\}$.

The rhombohedron, which we have already encountered as a special form in our description of the holosymmetric class of the trigonal system, is here a general form. The remarks made above concerning the apparent high symmetry of the general form in some classes clearly apply here, since a rhombohedron developed alone displays geometrically three planes of symmetry and three diad axes. We shall discuss later (p. 151) methods by which we might show that the underlying structure of such a rhombohedron in this class does not possess these symmetry elements, but meanwhile we may note that if two or more rhombohedra are developed together (Fig. 207) the absence of these elements of symmetry is clearly indicated.

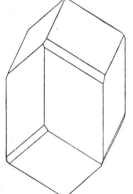

Fig. 207. A crystal of dioptase.

Examples. The substance usually regarded as the type of this class is the silicate of copper, H_2CuSiO_4, dioptase (Fig. 207) (though it is possible that the true symmetry of the structure is only that of class 3);

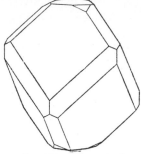

sodium sulphite, Na_2SO_3; the periodate, $(NH_4)_2H_3IO_6$ (Fig. 208); the lithium compounds, Li_2BeF_4, Li_2MoO_4, Li_2WO_4; and

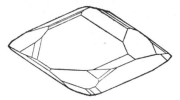

FIG. 208. A crystal of $(NH_4)_2H_3IO_6$.

FIG. 209. A crystal of phenacite.

amongst minerals (in addition to dioptase) dolomite, $CaMg(CO_3)_2$; willemite, Zn_2SiO_4; phenacite, Be_2SiO_4 (Fig. 209). The figure of this last substance illustrates a combination of eight different general forms.

CLASS 3m. (Ditrigonal hemimorphic, ditrigonal pyramidal.) Three vertical planes of symmetry intersecting in a triad axis (Fig. 210).

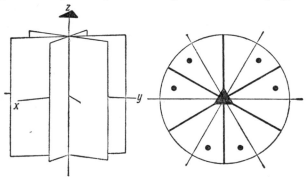

FIG. 210. Class 3m; the elements of symmetry, crystallographic axes and a stereogram of the general form.

		Miller-Bravais	Miller
Special forms.	Pedions	$\{0001\}$, $\{000\bar{1}\}$.	$\{111\}$, $\{\bar{1}\bar{1}\bar{1}\}$.
	Trigonal prisms	$\{10\bar{1}0\}$, $\{01\bar{1}0\}$.	$\{2\bar{1}\bar{1}\}$, $\{11\bar{2}\}$.
	Hexagonal prism	$\{11\bar{2}0\}$.	$\{10\bar{1}\}$.
	Ditrigonal prisms	$\{h\,k\,i\,0\}$.	$\{p\,q\,\overline{p+q}\}$.
	Trigonal pyramids	$\{h\,0\,\bar{h}\,l\}$,	$\{p\,q\,q\}$,
		$\{0\,k\,\bar{k}\,l\}$.	$\{p\,p\,q\}$.
	Hexagonal pyramids	$\{h\,h\,\overline{2h}\,l\}$.	$\{p\,q\,\overline{2q-p}\}$.
General forms.	Ditrigonal pyramids	$\{h\,k\,i\,l\}$.	$\{p\,q\,r\}$.

The triad axis in this class (as in class 3) is uniterminal; the presence of the three planes of symmetry introduces a number of special forms. The three kinds of prism should be noted; a prism face normal to a plane of symmetry, $10\bar{1}0$ or $01\bar{1}0$, is part of a *trigonal prism*, whilst the face $11\bar{2}0$, symmetrical to two planes of symmetry, is repeated six times to give a true hexagonal prism, but any other prism face $h\,k\,i\,0$ is repeated to give *three pairs* of faces. Such prisms (Fig. 211) are *ditrigonal prisms* in our nomenclature.

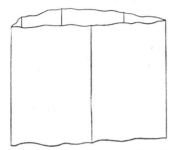

FIG. 211. Ditrigonal prism $\{h\,k\,i\,0\}$.

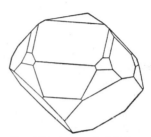

FIG. 212. A crystal of lithium sodium sulphate.

Examples. Lithium sodium sulphate, $LiNaSO_4$ (Fig. 212); the compound $LiNa_3(SO_4)_2 \,.\, 6H_2O$ and related chromates, selenates, molybdates and tungstates; potassium bromate, $KBrO_3$. Mineralogical examples are provided by the 'ruby-silvers'—pyrargyrite Ag_3SbS_3 and proustite Ag_3AsS_3—and by the important complex borosilicate, tourmaline, after which this class is often named. The crystal of tourmaline, illustrated in Fig. 213, shows a development of two trigonal pyramids and one ditrigonal pyramid at the upper end, the hexagonal prism and one trigonal prism, and one trigonal pyramid at the lower end, the uniterminal character of the triad axis being clearly evident.

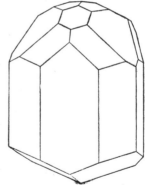

FIG. 213. A crystal of tourmaline.

CLASS $\bar{3}m$. (Trigonal holosymmetric, ditrigonal scalenohedral.) Three vertical planes of symmetry intersecting in an inversion triad axis (equivalent to three vertical planes of symmetry intersecting in a triad axis, three horizontal diad axes normal to the symmetry planes, and a centre), Fig. 214.

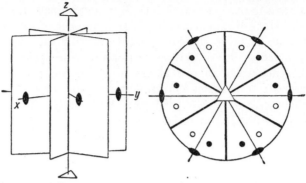

FIG. 214. Class $\bar{3}m$; the elements of symmetry, crystallographic axes and a stereogram of the general form.

Special forms.

	Miller-Bravais	*Miller*
Pinacoid	$\{0001\}$.	$\{111\}$.
Hexagonal prisms	$\{10\bar{1}0\}$, $\{11\bar{2}0\}$.	$\{2\bar{1}\bar{1}\}$, $\{10\bar{1}\}$.
Dihexagonal prisms	$\{h\,k\,i\,0\}$.	$\{p\,q\,\overline{-p+q}\}$.
Rhombohedra	$\{h\,0\,\bar{h}\,l\}$, $\{0\,k\,\bar{k}\,l\}$.	$\{p\,q\,q\}$, $\{p\,p\,q\}$.
Hexagonal bipyramids	$\{h\,h\,\overline{2h}\,l\}$.	$\{p\,q\,\overline{2q-p}\}$.

General Forms.

Ditrigonal scalenohedra $\{h\,k\,i\,l\}$. $\{p\,q\,r\}$.

Since the class $3/mm$ at the foot of the trigonal column in the chart (p. 106) is removed to the hexagonal system in virtue of its possession of an inverse hexad axis, the class $\bar{3}m$ under consideration ranks as the holosymmetric class of the trigonal system and as such has already been described (p. 84).

Examples. The type substance in this class is the trigonal modification of calcium carbonate, $CaCO_3$, the mineral calcite (Figs. 150, 161).

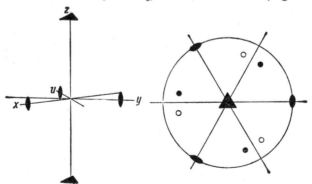

FIG. 215. Class 32; the elements of symmetry, crystallographic axes and a stereogram of the general form.

The isomorphous carbonates $FeCO_3$, $MgCO_3$, $MnCO_3$ and $ZnCO_3$ also belong here. Further examples are found in corundum, Al_2O_3 (Fig. 160); hematite, Fe_2O_3; sodium nitrate, $NaNO_3$; cadmium chloride, $CdCl_2$; magnesium hydroxide, $Mg(OH)_2$ (the mineral brucite).

CLASS 32. (Trigonal trapezohedral.) A triad axis normal to three diad axes (Fig. 215).

Special forms.	*Miller-Bravais*	*Miller*
Pinacoid	$\{0001\}$.	$\{111\}$.
Hexagonal prism	$\{10\bar{1}0\}$.	$\{2\bar{1}\bar{1}\}$.
Trigonal prisms	$\{2\bar{1}\bar{1}0\}$, $\{11\bar{2}0\}$.	$\{1\bar{1}0\}$, $\{10\bar{1}\}$.
Ditrigonal prisms	$\{h\,k\,i\,0\}$.	$\{p\,q\,-p+q\}$.
Rhombohedra	$\{h\,0\,\bar{h}\,l\}$, $\{0\,k\,\bar{k}\,l\}$.	$\{p\,q\,q\}$, $\{p\,p\,q\}$.
Trigonal bipyramids	$\{h\,h\,\overline{2h}\,l\}$,	$\{p\,q\,\overline{2q-p}\}$,
	$\{2h\,\bar{h}\,\bar{h}\,l\}$.	$\{p\,\overline{2q}-p\,q\}$.
General forms.		
Trigonal trapezohedra	$\{h\,k\,i\,l\}$.	$\{p\,q\,r\}$.

The general form in this class is a closed form composed of similar upper and lower portions (Fig. 216). It differs in an important respect from those forms which we have termed bi-pyramids, for there is no horizontal plane of symmetry. Each face is an irregular quadrilateral, and for such a form (we shall encounter related ones in the tetragonal and hexagonal systems) we shall use the term *trapezohedron*. (This name is also given, by some crystallographers, to the forms $\{h\,l\,l\}$ in the cubic system which we have called icositetrahedra (p. 53), but the faces of these forms have pairs of equal edges.) Corre-sponding to any one trapezohedron there can exist another, the faces of which are mirror-images of those of the former; the two forms are enantiomorphs, similar in shape but not superposable in space, and the class is another of

FIG. 216. Trigonal trapezohedron $\{h\,k\,i\,l\}$.

the important groups of symmetry displaying the phenomenon of enantiomorphism (p. 111). Fig. 217 illustrates a pair of trapezohedra orientated in such a way as to display this relationship clearly.

Examples. The best-known example in this class is the mineral quartz, the crystalline modification of SiO_2 stable at ordinary tempera-tures. Crystals of quartz frequently display special forms only—the hexagonal prism, rhombohedra, and sometimes a trigonal bipyramid—

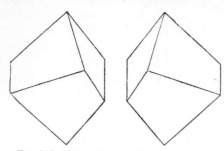

FIG. 217. Trigonal trapezohedra drawn in an orientation which displays clearly their enantiomorphous relationship.

but examples showing one or more trapezohedra are not unduly rare. The forms present on the crystal illustrated in Fig. 218 are usually indexed $\{10\bar{1}0\}$, $\{10\bar{1}1\}$, $\{01\bar{1}1\}$, $\{11\bar{2}1\}$, $\{51\bar{6}1\}$ (or, in Miller's notation, $\{2\bar{1}\bar{1}\}$, $\{100\}$, $\{22\bar{1}\}$, $\{41\bar{2}\}$, $\{4\bar{1}\bar{2}\}$); the crystal of Fig. 219 is the enantiomorph, showing the forms $\{10\bar{1}0\}$, $\{10\bar{1}1\}$, $\{01\bar{1}1\}$, $\{2\bar{1}\bar{1}1\}$, $\{6\bar{1}\bar{5}1\}$. Other substances belonging here are potassium dithionate, $K_2S_2O_6$, and the similar rubidium compound; lead dithionate,

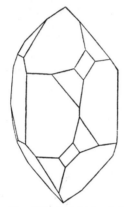

FIG. 218. A right-handed crystal of quartz.

FIG. 219. A left-handed crystal of quartz.

$PbS_2O_6 . 4H_2O$ (Fig. 220), and the similar compounds of calcium and strontium; rubidium tartrate, $Rb_2C_4H_4O_6$; and, as a further mineralogical example, cinnabar, HgS (Fig. 221).

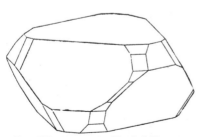

FIG. 220. A crystal of lead dithionate.

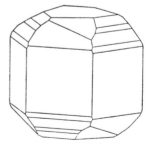

FIG. 221. A crystal of cinnabar.

TETRAGONAL SYSTEM

CLASS 4. (Tetragonal hemimorphic, tetragonal pyramidal.) A single tetrad axis (Fig. 222).

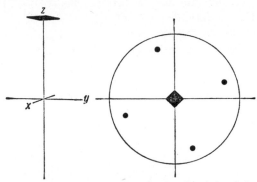

FIG. 222. Class 4; the crystallographic axes, showing their relationship to the tetrad axis, and a stereogram of the general form.

Special forms.   Pedions {001}, {00$\bar{1}$}.
                      Tetragonal prisms {$h\ k\ 0$}.
General forms.   Tetragonal pyramids {$h\ k\ l$}.

There is a close analogy between many of the classes of the tetragonal system and the related classes of the trigonal system. Thus class 4 resembles class 3, except that the rhythm is fourfold instead of three-fold. The general form is an open tetragonal pyramid, and the true

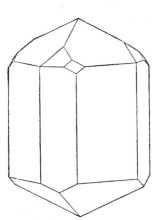

FIG. 223. A crystal of barium antimonyl tartrate.

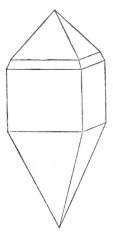

FIG. 224. A crystal of iodosuccinimide.

lack of symmetry may not be revealed by the crystallographic development, unless a suitable combination of forms is present. The tetrad
axis is uniterminal.

Examples.   Barium antimonyl tartrate, $Ba(SbO)_2(C_4H_4O_6)_2 . H_2O$
(Fig. 223); iodosuccinimide, $(CH_2CO)_2NI$ (Fig. 224); metaldehyde,
CH_3CHO; but the placing here is based in all three instances on other
considerations than the morphological development (see p. 151). The

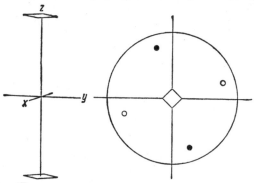

FIG. 225. Class $\bar{4}$; the crystallographic axes (the inversion tetrad axis flagged by
open squares) and a stereogram of the general form.

mineral wulfenite, $PbMoO_4$, was formerly assigned to this class but it
has now been shown to possess higher symmetry.

CLASS $\bar{4}$. (Tetragonal sphenoidal.)   A single inversion tetrad axis
(Fig. 225).

Special forms.   Pinacoid {001}.
                     Tetragonal prisms {$h\,k\,0$}.
General forms.   Tetragonal sphenoids {$h\,k\,l$}.

The general form in this class is a closed four-faced wedge-shaped

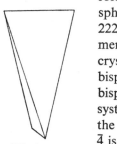

form (Fig. 226), which we shall call a (tetragonal)
sphenoid by analogy with the similar forms in class
222 of the orthorhombic system (p. 115).   As was
mentioned in the description of the latter class, some
crystallographers call a closed form of this kind a
bisphenoid or disphenoid, and to them class $\bar{4}$ is the
bisphenoidal (or disphenoidal) class of the tetragonal
system.   Since we are using such prefixes to denote
the presence of further elements of symmetry, class
$\bar{4}$ is in our nomenclature the *sphenoidal class* of the

FIG. 226. A tetragonal
sphenoid {$h\,k\,l$}.

tetragonal system.   The difference between a tetragonal sphenoid and a form of rather similar appearance in the orthorhombic system lies in the isosceles triangular shape of the faces of the former in contrast with the scalene triangular faces of the latter. As in many other classes of low symmetry, a single general form developed alone appears to possess vertical planes of symmetry, but this appearance of higher symmetry is no longer true of suitable combinations of forms; Fig. 227 represents a combination of three sphenoids, and the absence of any symmetry other than an inversion tetrad axis is clear.

Examples. This class of symmetry was formerly devoid of any satisfactory examples. A synthetic calcium aluminium silicate, quoted in

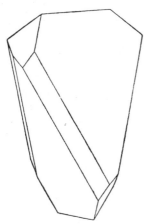

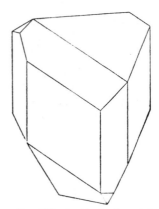

FIG. 227. A combination of three tetragonal sphenoids.

FIG. 228. An idealised crystal of cahnite.

most text-books, appears to have been a member of the melilite series of minerals and as such belongs to a higher class. Pentaerythritol, $C(CH_2OH)_4$, is placed here on other grounds than morphological development; other examples are probably to be found in boron phosphate, BPO_4, and the corresponding arsenate $BAsO_4$. An excellent mineralogical example is afforded by a recently-described mineral cahnite, $Ca_4B_2As_2O_{12} \cdot 4H_2O$; Fig. 228 is an idealised representation of a crystal of this substance, the natural crystals so far found being well terminated at one extremity only. Schreibersite, Fe_3P, may also belong to this class.

CLASS $4/m$. (Tetragonal bipyramidal.) A tetrad axis normal to a plane of symmetry, and a centre (Fig. 229).

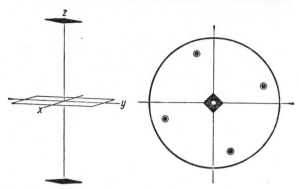

FIG. 229. Class $4/m$; the elements of symmetry, crystallographic axes and a stereogram of the general form.

Special forms.   Pinacoid {001}.
    Tetragonal prisms {$h\,k\,0$}.
General forms.   Tetragonal bipyramids {$h\,k\,l$}.

Examples.   The tungstates and molybdates of calcium, barium and lead afford characteristic examples.   Fig. 230 illustrates a crystal of scheelite, $CaWO_4$, showing four tetragonal bipyramids.   The crystal of wulfenite, $PbMoO_4$ (Fig. 231), shows how the combination of a prism

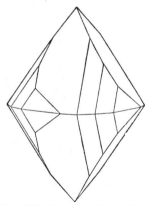

FIG. 230. A crystal of scheelite.   FIG. 231. A crystal of wulfenite.

with a single bipyramid may reveal clearly the absence of vertical planes of symmetry.   Further examples are found in the anhydrous periodates $NaIO_4$, KIO_4.

CLASS $4mm$.   (Ditetragonal hemimorphic, ditetragonal pyramidal.)
    Two pairs of planes of symmetry at right-angles intersecting in a single tetrad axis (Fig. 232).

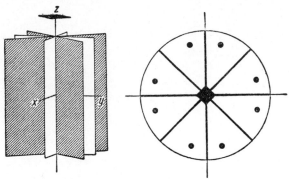

FIG. 232. Class 4*mm*; the elements of symmetry (with the two pairs of planes of symmetry differently shaded), crystallographic axes and a stereogram of the general form.

Special forms.   Pedions {001}, {00$\bar{1}$}.
Tetragonal prisms {100}, {110}.
Ditetragonal prisms {*h k* 0}.
Tetragonal pyramids {*h* 0 *l*}, {*h h l*}.

General forms.   Ditetragonal pyramids {*h k l*}.

This class of the tetragonal system may be compared with class 3*m*, the ditrigonal pyramidal class of the trigonal system. It will be noticed, however, that whilst the three planes of symmetry in class 3*m* are all similar planes, related to each other by the triad axis, the planes of symmetry in class 4*mm* belong to *two pairs*, differently shaded in Fig. 232. The first *m* of the symbol refers to the pair normal to the directions chosen as crystallographic *x* and *y*; the second *m* introduced into the symbol refers to the pair of planes set diagonally to these.

Examples. The general form of this class has probably never yet been observed on actual crystals, so that the allocation of any substance to this class must have involved other considerations to be discussed later. The best example is provided by the mineral diaboleite, $Pb_2CuCl_2(OH)_4$, of which a recently-described example is illustrated in Fig. 233.

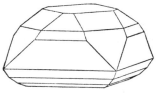

FIG. 233. A crystal of diaboleite.

CLASS $\bar{4}$2*m*. (Tetragonal bisphenoidal, tetragonal scalenohedral.) Two planes of symmetry at right angles intersecting in an inversion tetrad axis, and two diad axes at 45° to the planes (Fig. 234). The diad axes are chosen as *x* and *y* axes (see Chart, p. 106).

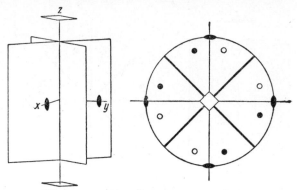

FIG. 234. Class $\bar{4}2m$; the elements of symmetry, crystallographic axes and a stereogram of the general form.

Special forms. Pinacoid {001}.
Tetragonal prisms {100}, {110}.
Ditetragonal prisms {$h\,k\,0$}.
Tetragonal bipyramids {$h\,0\,l$}.
Sphenoids {$h\,h\,l$}, {$h\,\bar{h}\,l$}.

General forms. Bisphenoids {$h\,k\,l$}.

In this class the sphenoids are all special forms, developed from faces $h\,h\,l$ or $h\,\bar{h}\,l$ normal to a plane of symmetry. Fig. 235 illustrates a sphenoid {$h\,h\,l$}, which thus shows (structurally as well as morpho-

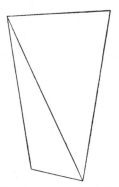

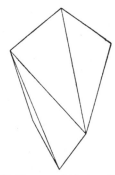

FIG. 235. Tetragonal sphenoid {$h\,h\,l$}.   FIG. 236. Tetragonal bisphenoid {$h\,k\,l$}.

logically) two planes of symmetry. The general form consists of four pairs of faces (Fig. 236), since a generally-situated face $h\,k\,l$ is reflected across both planes of symmetry. Consistently with our usage of the prefix *bi*-, we shall call such forms *bisphenoids*; those crystallographers who have already appropriated this name for the four-faced sphenoids use the alternative name *tetragonal scalenohedron*. (Considerable confusion has arisen in the past owing to this confused application of the terms sphenoid and bisphenoid (or disphenoid). One well-known text-book, for example, terms the class under consideration sphenoidal, though referring to the general form as a tetragonal scalenohedron, and applies the term tetragonal disphenoidal to class $\bar{4}$, a quite indefensible proceeding! The confusion will abate as the internationally-accepted symbolic notation becomes more widely used, but for us meanwhile class $\bar{4}$ will be the sphenoidal class of the tetragonal system, and class $\bar{4}2m$ the bisphenoidal class.)

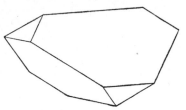

FIG. 237. A crystal of tetraethyl-ammonium iodide.

Examples. Tetraethylammonium iodide, $N(C_2H_5)_4I$ (Fig. 237); mercuric cyanide, $Hg(CN)_2$ (Fig. 238); urea, $CO(NH_2)_2$ (Fig. 239);

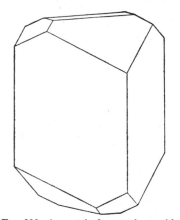

FIG. 238. A crystal of mercuric cyanide.

FIG. 239. A crystal of urea.

potassium dihydrogen phosphate, KH_2PO_4 and the isomorphous ammonium compound. Amongst minerals, the best example is afforded by chalcopyrite, $CuFeS_2$, after which the class is often named; Fig. 240 illustrates a typical sphenoidal habit, a combination of a sphenoid $\{h\,h\,l\}$ with a general form, whilst Fig. 241 displays a different habit owing to the predominance of bipyramidal special forms. Other examples are provided by the melilite group of silicates.

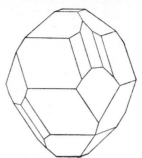

FIG. 240. A sphenoidal crystal of chalcopyrite.

FIG. 241. A chalcopyrite crystal of bipyramidal habit.

CLASS 42. (Tetragonal trapezohedral.) A tetrad axis normal to two pairs of diad axes (Fig. 242).

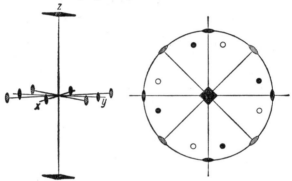

FIG. 242. Class 42; the elements of symmetry (the two pairs of diad axes have differently shaded flags), crystallographic axes and a stereogram of the general form.

Special forms.   Pinacoid {001}.
Tetragonal prisms {100}, {110}.
Ditetragonal prisms {h k 0}.
Tetragonal bipyramids {h 0 l}, {h h l}.

General forms.   Tetragonal trapezohedra {h k l}.

This class is the tetragonal analogue of the class 32 of the trigonal system, and the general form (Fig. 243) is of the type which we have agreed to call trapezohedral; corresponding to each trapezohedron {h k l} there can exist another, {k h l}, which is its enantiomorph. Notice that in this class, although, as in class 32, there is no centre of symmetry, the even degree of the vertical axis results in the horizontal diad axes not being uniterminal.

Examples.   The general form has not been very frequently observed. Of the few examples available we may mention methylammonium iodide,

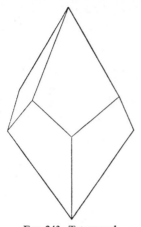

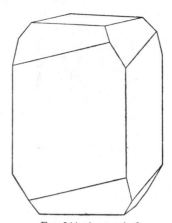

FIG. 243. Tetragonal
trapezohedron {*h k l*}.

FIG. 244. A crystal of
methylammonium iodide.

$NH_3(CH_3)I$ (Fig. 244); guanidine carbonate, $2CNH(NH_2)_2 . H_2CO_3$; and, as a mineralogical example, phosgenite, $(PbO)_2CCl_2O$ (Fig. 245). Further examples are found in substances placed here for reasons other than form development, and crystals show only special forms, as in nickel sulphate, $NiSO_4 . 6H_2O$ (Fig. 246); ethylene diamine sulphate, $C_2H_4(NH_2)_2 . H_2SO_4$; ammonium uranyl acetate, $NH_4UO_2(CH_3COO)_3$.

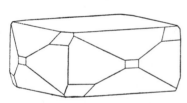

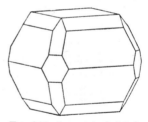

FIG. 245. A crystal of phosgenite.

FIG. 246. A crystal of nickel
sulphate.

CLASS 4/*mmm*. (Tetragonal holosymmetric, ditetragonal bipyramidal.) A tetrad axis at the intersection of two pairs of planes of symmetry, two pairs of horizontal diad axes normal to these planes, a plane of symmetry normal to the tetrad axis and a centre (Fig. 247).

Special forms.   Pinacoid {001}.
Tetragonal prisms {100}, {110}.
Ditetragonal prisms {*h k* 0}.
Tetragonal bipyramids {*h* 0 *l*}, {*h h l*}.

General forms.   Ditetragonal bipyramids {*h k l*}.

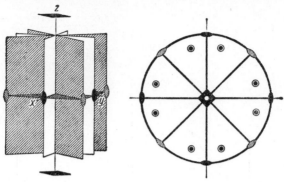

FIG. 247. Class 4/*mmm*; the elements of symmetry, crystallographic axes and a stereogram of the general form.

Examples. We have already illustrated tetramethylammonium iodide (Fig. 114) and mercurous chloride (Fig. 115). As further examples may be mentioned cassiterite, SnO_2; rutile and anatase, both of composition TiO_2; zircon, $ZrSiO_4$; vesuvianite, a complex calcium aluminium silicate.

HEXAGONAL SYSTEM

CLASS 6. (Hexagonal hemimorphic, hexagonal pyramidal.) A single hexad axis (Fig. 248).

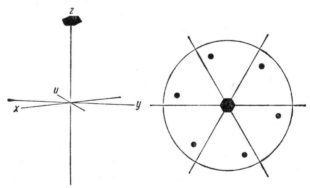

FIG. 248. Class 6; the crystallographic axes and a stereogram of the general form.

Special forms.   Pedions {0001}, {000$\bar{1}$}.
Hexagonal prisms {$h\,k\,i\,0$}.
General forms.   Hexagonal pyramids {$h\,k\,i\,l$}.

The hexad axis is clearly uniterminal; the class should be compared with class 3 and class 4.

Examples. This is a further class in which the necessary combination of forms to enable one to place a substance indubitably in the class on morphological grounds is rarely found. Crystals are usually of simple habit, and are placed here on evidence other than that of form development. We may mention lithium potassium

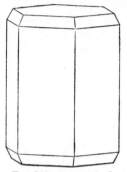

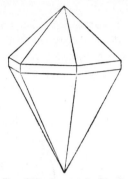

FIG. 249 .A crystal of lithium potassium sulphate.

FIG. 250. A crystal of lead antimonyl tartrate.

FIG. 251. A crystal of nepheline.

sulphate, $LiKSO_4$ (Fig. 249); iodoform, CHI_3; lead antimonyl tartrate, $Pb(SbO)_2(C_4H_4O_6)_2$ (Fig. 250); nepheline, $NaAlSiO_4$ (Fig. 251).

CLASS $\bar{6}$. (Trigonal bipyramidal.) An inversion hexad axis (equivalent to a triad axis normal to a plane of symmetry, $\bar{6} = 3/m$, Fig. 252).

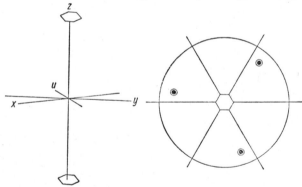

FIG. 252. Class $\bar{6}$; the inversion axis (flagged with open hexagons), crystallographic axis and a stereogram of the general form.

Special forms.   Pinacoid {0001}.
Trigonal prisms {h k i 0}.

General forms.   Trigonal bipyramids {h k i l}.

This class is one of two formerly placed in the trigonal system but now allocated more appropriately to the hexagonal system. The vertical axis is morphologically a triad axis, and the adjective *trigonal* must still be used in describing the prisms and bipyramids. Structurally, however, a crystal belonging to this class would show a more symmetrical arrangement around this axis than that necessitated by a rotary triad axis—the arrangement consistent with an inversion hexad axis.

Examples. Only one substance, the unstable silver phosphate Ag_2HPO_4, has been tentatively assigned to this class, but the evidence is insufficient, and we must regard class $\bar{6}$ as at present without an established representative amongst known crystalline substances.

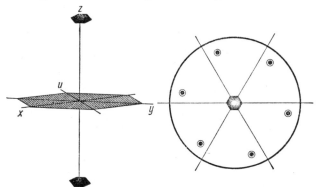

Fig. 253. Class $6/m$; the elements of symmetry, crystallographic axes and a stereogram of the general form.

CLASS $6/m$. (Hexagonal bipyramidal.) A hexad axis normal to a plane of symmetry, and a centre (Fig. 253).

Special forms.   Pinacoid $\{0001\}$.
            Hexagonal prisms $\{h\,k\,i\,0\}$.
General forms.   Hexagonal bipyramids $\{h\,k\,i\,l\}$.

Examples. This is yet another class with few known representatives,

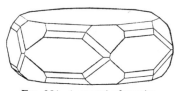

Fig. 254. A crystal of apatite.

and the best examples are found in the apatite group of minerals. Fig. 254 represents a crystal of apatite itself, $(CaF)Ca_4(PO_4)_3$; the development of one example of the general form shows clearly the presence of a horizontal plane of symmetry and the lack of any vertical planes of symmetry.

CLASS $6mm$. (Dihexagonal hemimorphic, dihexagonal pyramidal.) A hexad axis at the intersection of two sets of three planes of symmetry (Fig. 255).

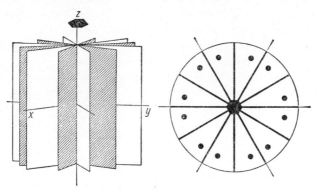

FIG. 255. Class 6*mm*; the elements of symmetry (the two families of planes of symmetry are differently shaded), crystallographic axes and a stereogram of the general form.

Special forms.   Pedions {0001}, {000$\bar{1}$}.

Hexagonal prisms {10$\bar{1}$0}, {11$\bar{2}$0}.

Dihexagonal prisms {$h\,k\,i\,0$}.

Hexagonal pyramids {$h\,0\,\bar{h}\,l$}, {$h\,h\,\overline{2h}\,l$}.

General forms.   Dihexagonal pyramids {$h\,k\,i\,l$}.

This class of the hexagonal system should be compared with class 3*m* of the trigonal system and with class 4*mm* of the tetragonal system. It is a further class with a clearly uniterminal principal axis.

Examples. The general form in this class has rarely been observed, a common type of habit being that of the crystal of zincite, ZnO, illustrated in Fig. 256. Fig. 257 illustrates a recently-described specimen of artificially-crystallised ZnO in which a general form, indexed {21$\bar{3}\bar{3}$}, appears on the lower portion of the crystal; the other forms present are the pedion {0001} small, the pedion {000$\bar{1}$} large, hexagonal prism {10$\bar{1}$0}, hexagonal pyramids {10$\bar{1}$1} and {11$\bar{2}$2}. Other examples of this

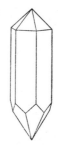

FIG. 256. A crystal of zincite.

FIG. 257. A more highly modified crystal of zincite.

FIG. 258. A crystal of triethylammonium chloride.

symmetry are found in greenockite, CdS; wurtzite, ZnS; iodyrite, AgI; bromellite, BeO; triethylammonium chloride, $NH(C_2H_5)_3Cl$ (Fig. 258).

CLASS $\bar{6}m2$. (Ditrigonal bipyramidal.) An inversion hexad axis at the intersection of three vertical planes of symmetry (equivalent to a triad axis normal to a plane of symmetry, three planes of symmetry intersecting in the triad axis, three diad axes lying in these planes normal to the triad axis, $\bar{6}m = 3/mm$, Fig. 259).

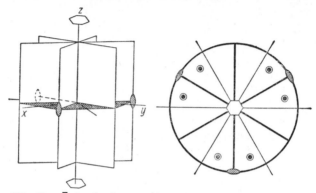

FIG. 259. Class $\bar{6}m2$; the elements of symmetry, crystallographic axes and **a** stereogram of the general form.

Special forms.   Pinacoid {0001}.
Trigonal prisms {10$\bar{1}$0}, {01$\bar{1}$0}.
Hexagonal prism {11$\bar{2}$0}.
Ditrigonal prisms {$h\,k\,i\,0$}.
Trigonal bipyramids {$h\,0\,\bar{h}\,l$}, {$0\,k\,\bar{k}\,l$}.
Hexagonal bipyramids {$h\,h\,\overline{2h}\,l$}.

General forms.   Ditrigonal bipyramids {$h\,k\,i\,l$}.

This is the second class in which an axis which is morphologically one of threefold symmetry has structurally the higher symmetry $\bar{6}$, and the class is therefore placed in the hexagonal system.

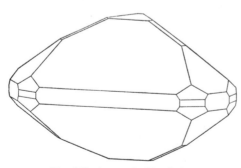

FIG. 260. A crystal of benitoite.

Examples. The only certain example of this class of symmetry is found in a rare mineral, benitoite, $BaTiSi_3O_9$. Usually only special forms are present

(Fig. 260), but crystals have also been described showing rather dubious examples of a general form (Fig. 261). Notice that these figures are orientated so that the crystallographic axes x, y and u are chosen normal to the planes of symmetry, and therefore not parallel to the diad axes; it is for this reason that the symbol of the class is written $\bar{6}m2$, and not merely $\bar{6}m$, but we shall explain the significance of this introduction more fully when we come to study the symmetry of the internal structural pattern in such crystals (p. 256).

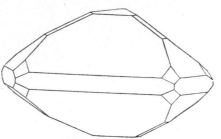

FIG. 261. A crystal of benitoite showing a general form.

CLASS 62. (Hexagonal trapezohedral.) A hexad axis normal to six diad axes (Fig. 262).

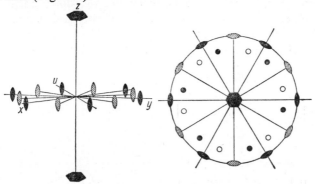

FIG. 262. Class 62; the elements of symmetry, crystallographic axes and stereogram of the general form.

Special forms.	Pinacoid {0001}.

Hexagonal prisms {10$\bar{1}$0}, {11$\bar{2}$0}.
Dihexagonal prisms {$h\,k\,i\,0$}.
Hexagonal bipyramids {$h\,0\,\bar{h}\,l$}, {$hh\,\overline{2h}\,l$}.

General forms. Hexagonal trapezohedra {$h\,k\,i\,l$}.

This class is to be compared with class 32 of the trigonal system and class 42 of the tetragonal system. The general form is such that a form {$h\,k\,i\,l$} is the enantiomorph of the form {$i\,\bar{k}\,\bar{h}\,l$} (Fig. 263), but such forms have apparently never yet been observed on actual crystals.

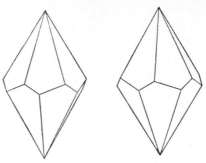

FIG. 263. An enantiomorphous pair of hexagonal trapezohedra.

Examples.   Since the general form has never been observed, other considerations have been used to place in this class lithium iodate, $LiIO_3$; barium aluminate, $BaAl_2O_4$; high-quartz (the modification of crystalline SiO_2 stable between 573° C. and 870° C.); dibenzalpentaerythritol; kalsilite, $KAlSiO_4$; taaffeite, $BeMgAl_4O_8$.

CLASS 6/*mmm*.   (Hexagonal holosymmetric, dihexagonal bipyramidal.)   A hexad axis at the intersection of two sets of three vertical planes of symmetry, two sets of three diad axes normal to these planes, a plane of symmetry normal to the hexad axis, and a centre (Fig. 264).

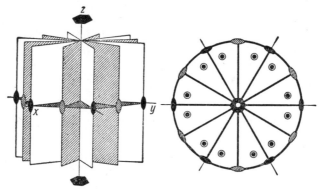

FIG. 264. Class 6/*mmm*; the elements of symmetry, crystallographic axes and a stereogram of the general form.

Special forms.   Pinacoid {0001}.

Hexagonal prisms {10$\bar{1}$0}, {11$\bar{2}$0}.

Dihexagonal prisms {$h k i$ 0}.

Hexagonal bipyramids {h 0 $\bar{h}$ l}, {$h h \overline{2h} l$}.

General forms.   Dihexagonal bipyramids {$h k i l$}.

This is the holosymmetric class of the hexagonal system, which we have already studied. The third *m* of the symbol refers to the second set of three planes of symmetry which arises automatically from the operations indicated by the symbol 6/*mm*.

Examples. The best example is afforded by the mineral beryl, $Be_3Al_2Si_6O_{18}$, a crystal of which is illustrated in Fig. 149 (p. 84).

CUBIC SYSTEM

In this system are placed all those groups of symmetry elements (crystal classes) which contain four triad axes. These axes, however, cannot exist alone. By inserting on a stereogram poles, in a general position, to satisfy the four triad axes (compare Fig. 266) the student can show that three diad axes at right angles are automatically introduced; thus the class of lowest symmetry in the cubic system possesses four triad axes and three diad axes.

In the conventional setting of a cube with one set of edges vertical the triad axes, the diagonals of the cube, occupy an inclined position; this is denoted in the symbol of each class by the appearance of the figure 3 in a position other than at the beginning, such as 23. A special convention must be observed, too, in the use of the symbol *m* to denote the presence of a plane of symmetry, for we have seen already that there may be present planes parallel to the cube faces, planes parallel to the dodecahedral faces, or both kinds of symmetry plane together. The former type, passing through the vertical principal axis of symmetry but not through a triad axis, is symbolised by a letter *m*

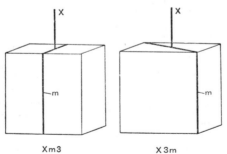

FIG. 265. The symbolisation of the two kinds of plane of symmetry possible in cubic crystals; the figures show only one plane of each family.

preceding the figure 3, as *Xm*3; the latter type passes also through a triad axis, and its symbol is therefore placed after the figure 3, as *X*3*m* (Fig. 265). The vertical axis may be a diad axis, an inversion tetrad axis or a rotation tetrad axis, and trial will show that the five symbols in the first column below represent all the different combinations possible:

Full symbol	Abbreviated symbol
23	23
432	43
2m3	m3
4̄3m	4̄3m
4m3m2	m3m

These symbols are conventionally abbreviated to those in the second column by omitting symmetry elements which arise automatically from a given combination. Thus in the class with four triad axes and symmetry planes parallel to the cube faces, diad axes normal to the cube faces will inevitably arise, so that the full symbol 2m3 is sufficiently indicated by the abbreviated version m3.

CLASS 23. (Tetrahedral pentagonal dodecahedral.) Three mutually perpendicular diad axes and four triad axes (Fig. 266).

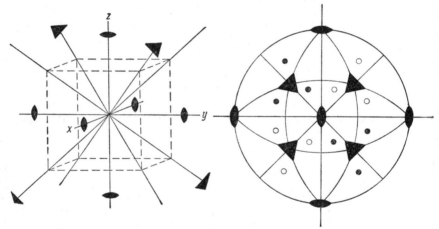

Fig. 266. Class 23; the elements of symmetry, crystallographic axes and a stereogram of the general form. The outline of the cube has been added to the drawing of the elements of symmetry to help in visualising their arrangement in space.

Special forms.   Cube {100}.
Rhombic dodecahedron {110}.
Pentagonal dodecahedra {h k 0}, {k h 0}.
Tetrahedra {111}, {11̄1}.
Tristetrahedra {h l l}, {h l̄ l}.
Deltoid dodecahedra {h h l}, {h h̄ l}.

General forms.   Tetrahedral pentagonal dodecahedra {h k l}.

The first four types of special form we have already described in our preliminary study of the cubic system; it may be observed that the appearance of pentagonal dodecahedra as special forms in this class supports our objection to the alternative name *pyritohedra* (p. 60), for this is not the class to which the mineral pyrite belongs. The *tristetrahedra* (sometimes, like other names of this kind, written in full as triakistetrahedra) belong to two families, $\{h\,l\,l\}$ and $\{h\,\bar{l}\,l\}$, of which the representatives $\{211\}$ and $\{2\bar{1}1\}$ are illustrated in Figs. 267 and 268; the

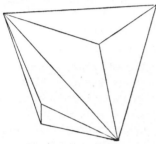

FIG. 267. The tristetrahedron $\{211\}$.

FIG. 268. The tristetrahedron $\{2\bar{1}1\}$.

two families are distinguished by some authors as 'positive' and 'negative' tristetrahedra respectively. Fig. 269 illustrates a further member, $\{311\}$. The *deltoid dodecahedra* similarly occur in two families $\{h\,h\,l\}$ and $\{h\,\bar{h}\,l\}$, of which $\{221\}$ and $\{2\bar{2}1\}$ are illustrated in Figs. 270 and 271. Notice that neither of these pairs of families displays enantiomorphism, for a 'negative' form can be brought into congruence with

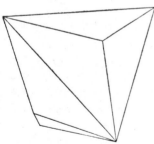

FIG. 269. The tristetra-
hedron $\{311\}$.

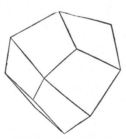

FIG. 270. The deltoid
dodecahedron $\{221\}$.

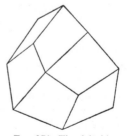

FIG. 271. The deltoid
dodecahedron $\{2\bar{2}1\}$.

the corresponding 'positive' form by a rotation of 90° about the vertical axis. The general forms (the most appropriate name for these seems to be the lengthy but descriptive *tetrahedral pentagonal dodecahedron*) do display enantiomorphism, however, since corresponding to every form {h k l} there can exist a form {k h l} which is its mirror image. Fig. 272 illustrates the form {321} and Fig. 273 its enantiomorph {231}, whilst the enantiomorphous pair {3$\bar{2}$1} and {2$\bar{3}$1} are shown in Figs. 274, 275. In conclusion, we may note that the triad axes in this class are uniterminal (Fig. 266).

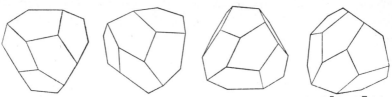

FIGS. 272-275. Tetrahedral pentagonal dodecahedra {321}   {231}   {3$\bar{2}$1}   {2$\bar{3}$1}.

Examples. The classic examples of this class of symmetry are provided by sodium chlorate, $NaClO_3$, and sodium bromate, $NaBrO_3$. Figs. 276, 277 represent the habits of crystals obtained from aqueous solution at ordinary temperatures; as usually described, the forms represented are the cube {100}, rhombic dodecahedron {110} and

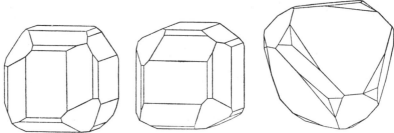

FIG. 276. A crystal of sodium chlorate.

FIG. 277. A crystal of sodium chlorate which is morphologically the enantiomorph of the crystal in Fig. 276.

FIG. 278. A crystal showing a tetrahedral pentagonal dodecahedron in combination with special forms.

tetrahedron {1$\bar{1}$1}, with the pentagonal dodecahedron {210} in Fig. 276 and the pentagonal dodecahedron {120} in Fig. 277, but the two combinations are morphologically enantiomorphs. The general form has rarely been observed, but Fig. 278 illustrates the ideal symmetrical development of a crystal of tetrahedral habit in which a tetrahedral pentagonal dodecahedron is combined with the tetrahedron {111}, tetrahedron {1$\bar{1}$1}, cube {100} and rhombic dodecahedron {110}. Further examples of this class of symmetry are found in Schlippe salt, $Na_3SbS_4 . 9H_2O$; sodium calcium silicate, Na_2CaSiO_4; sodium uranyl acetate, $NaUO_2(CH_3COO)_3$; and the minerals ullmannite, $NiSbS$, and cobaltite, $CoAsS$. The nitrates of strontium, barium and lead were formerly placed here but are now assigned to class $m3$ (below).

CLASS *m*3. (Di(akis)dodecahedral.) Three diad axes at the inter-sections of three mutually perpendicular planes of symmetry, four triad axes and a centre (Fig. 279).

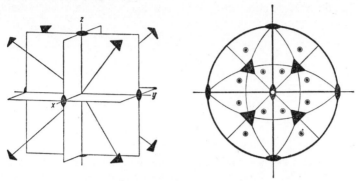

FIG. 279. Class *m*3; the elements of symmetry, crystallographic axes and a stereogram of the general form.

Special forms.   Cube {100}.
Rhombic dodecahedron {110}.
Pentagonal dodecahedra {*h k* 0}, {*k h* 0}.
Octahedron {111}.
Icositetrahedra {*h l l*}.
Trisoctahedra {*h h l*}.

General forms.   Di(akis)dodecahedra {*h k l*}.

The general form, which we shall term a *didodecahedron*, is some-times called a *diploid*. Corresponding to a form {*h k l*} there is a form {*h l k*} of identical shape, but with a different attitude (and hence a different relationship to the underlying structure). Fig. 280 illustrates the didodecahedron {321}, and Fig. 281 the form {312}. The form {421}, a further member of the family, is portrayed in Fig. 282, and it will be seen to present a particular peculiarity; a pair of opposite edges of every face are parallel, so that the form shows prominent zonal relation-

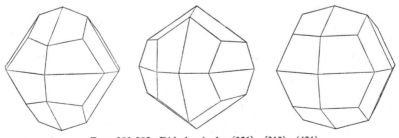

FIGS. 280-282. Didodecahedra {321}   {312}   {421}.

ships which are not characteristic of didodecahedra in general. The combination of the cube {100} with this form (Fig. 283) is of some interest, since it appears at first sight to consist of thirty similar rhombus faces, and therefore apparently to possess axes of pentagonal symmetry.

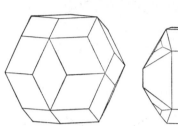  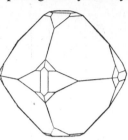

FIG. 283. The didodecahedron {421} modified by faces of the cube.

FIG. 284. A crystal of pyrite.

FIG. 285. A crystal of tin iodide.

(Parallelism of edges occurs in a didodecahedron $\{h\,k\,l\}$ if $k^2 = hl$, so that {421} is the only example likely to be encountered in practice.)

Examples.   The type example is found in the mineral pyrite (iron pyrites), FeS_2.   Fig. 284 illustrates a crystal showing the cube {100}, pentagonal dodecahedron {210}, octahedron {111} and one example of the general form.   The crystal of tin iodide, SnI_4, in Fig. 285, displays only special forms, but the twofold symmetry of the crystallographic axes is clearly evident.   Also placed in this class are the alums, $R'R'''(SO_4)_2 . 12H_2O$;   the   nitrates $Ca(NO_3)_2$, $Sr(NO_3)_2$, $Ba(NO_3)_2$,

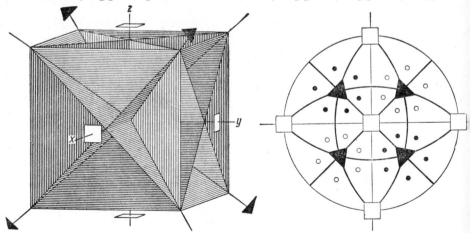

FIG. 286. Class $\bar{4}3m$;   the elements of symmetry, crystallographic axes and a stereogram of the general form.

$Pb(NO_3)_2$; zinc bromate, $Zn(BrO_3)_2 . 6H_2O$. Further mineralogical examples are provided by hauerite, MnS_2; sperrylite, $PtAs_2$; bixbyite, $(Fe,Mn)_2O_3$.

CLASS $\bar{4}3m$. (Hexa(kis)tetrahedral.) Three diad axes (actually these are inversion tetrad axes), four triad axes and six planes of symmetry (Fig. 286).

Special forms.   Cube {100}.
Rhombic dodecahedron {110}.
Tetrahexahedra {$h\,k\,0$}.
Tetrahedra {111}, {1$\bar{1}$1}.
Tristetrahedra {$h\,l\,l$}, {$h\,\bar{l}\,l$}.
Deltoid dodecahedra {$h\,h\,l$}, {$h\,\bar{h}\,l$}.
General forms.   Hexa(kis)tetrahedra {$h\,k\,l$}, {$h\,\bar{k}\,l$}.

Corresponding to each *hexatetrahedron* {$h\,k\,l$} (Fig. 287) there is a possible related hexatetrahedron {$h\,\bar{k}\,l$} (Fig. 288), but owing to the

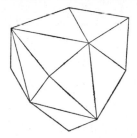

FIG. 287. The hexatetrahedron {321}.   FIG. 288. The hexatetrahedron {3$\bar{2}$1}.

presence of planes of symmetry in this class the two forms are morphologically identical and are not enantiomorphs. The triad axes in this class are clearly uniterminal. The general appearance of crystals possessing this symmetry varies considerably according to the pre-

FIG. 289. A combination of the rhombic dodecahedron {110} with the tristetrahedron {311}.

FIG. 290. A combination of the forms {110}, {111}, {100} and {311}.

dominance of 'tetrahedral' forms (the tetrahedra, tristetrahedra, deltoid dodecahedra and hexatetrahedra) or of the apparently holosymmetric special forms (the cube and rhombic dodecahedron). Fig. 289 shows a combination of the forms {110}, {311}; whilst Fig. 290 illustrates a more highly modified crystal in which {100} and {111} are also present.

Examples. The type substance in this class is the mineral zinc blende (sphalerite), ZnS, the commonest representative of a group of sulphides, selenides and tellurides of beryllium, zinc, cadmium and mercury belonging here (though a number of these substances are dimorphous, and also occur in modifications with the symmetry of class $6mm$). Also placed here are the cuprous halides CuCl, CuBr, CuI; aluminium metaphos-

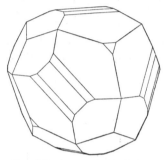

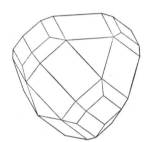

FIG. 291. A crystal of eulytine.   FIG. 292. A crystal of fahlerz.

phate, $Al(PO_3)_3$; silver phosphate, Ag_3PO_4; eulytine, $Bi_4(SiO_4)_3$ (Fig. 291); sulvanite, Cu_3VS_4; the fahlerz series between tetrahedrite, Cu_3SbS_3, and tennantite, Cu_3AsS_3 (Fig. 292); zunyite, $Al_{13}Si_5O_{20}(OH,F)_{18}Cl$.

CLASS 43. (Pentagonal icositetrahedral.)   Three tetrad axes, four triad axes and six diad axes (Fig. 293).

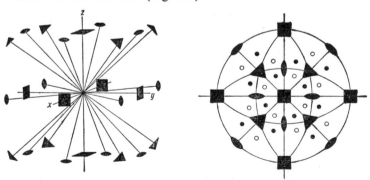

FIG. 293. Class 43; the elements of symmetry, crystallographic axes and a stereogram of the general form.

Special forms.   Cube {100}.
Rhombic dodecahedron {110}.
Tetrahexahedra {*h k* 0}.
Octahedron {111}.
Icositetrahedra {*h l l*}.
Trisoctahedra {*h h l*}.

General forms.   Pentagonal icositetrahedra {*h k l*}, {*k h l*}.

In the absence of planes of symmetry and of the operation of inversion, the general forms of this class (*pentagonal icositetrahedra*, sometimes called *gyroids*) can exist as true enantiomorphs (Figs. 294, 295).

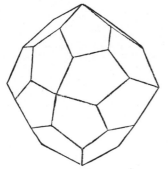 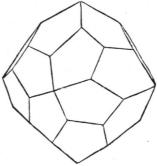

FIGS. 294-295. Pentagonal icositetrahedra {321} {231}.

The class lacks a centre of symmetry, but the triad axes are not uni-terminal, and the special forms are all identical with those of the holosymmetric class.

Examples.   This is one of the two possible classes (the other being $\bar{6}$, p. 136) for which no actual representative amongst known crystalline substances has yet been clearly established. Cuprous oxide, Cu_2O (cuprite), was formerly placed here on account of the supposed development of faces of pentagonal icositetrahedra (Fig. 296); the indices of these doubtful forms, however, were high ({968} and {13, 10, 12} in specimens from two different localities), and cuprite is now believed to be holosymmetric.

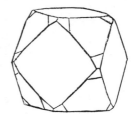

FIG. 296. A cube modified by the pentagonal icositetrahedron {968}

CLASS *m3m*. (Cubic holosymmetric, hex(akis)octahedral.)   Three tetrad axes, four triad axes, six diad axes, three cubic planes of symmetry, six dodecahedral planes of symmetry and a centre (Fig. 297).

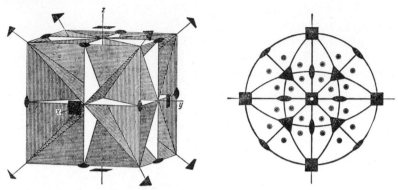

FIG. 297. Class $m3m$; the elements of symmetry, crystallographic axes and a stereogram of the general form.

Special forms.   Cube {100}.

Rhombic dodecahedron {110}.

Tetrahexahedra {$h\,k\,0$}.

Octahedron {111}.

Icositetrahedra {$h\,l\,l$}.

Trisoctahedra {$h\,h\,l$}.

General forms.   Hex(akis)octahedra {$h\,k\,l$}.

Examples. Many metals—such as copper, silver, gold, lead, platinum and iron—belong here. Fig. 298 illustrates a crystal of silver

showing a hexoctahedron {751} predominant, together with {211}; most of the other metals mentioned usually show only special forms. Further examples are provided by the halides of the alkali-metals, and the oxides, RO, sulphides, RS, selenides, RSe, and tellurides, RTe, of calcium, strontium, barium and lead. This series provides mineralogical examples in halite, NaCl; periclase, MgO; galena, PbS; alabandite, MnS; clausthalite, PbSe; altaite,

FIG. 298. A crystal of silver.

PbTe. We may mention also fluorite, CaF_2; the spinels, $R''R_2'''O_4$ (including magnetite, Fe_3O_4, Fig. 92); garnets, $R_3''R_2'''(SiO_4)_3$.

TABLE OF FORMS IN THE CUBIC SYSTEM

	23	$m3$	$\bar{4}3m$	43	$m3m$
Special forms {100} {110}	Cube Rhombic dodecahedron	Cube Rhombic dodecahedron	Cube Rhombic dodecahedron	Cube Rhombic dodecahedron	Cube Rhombic dodecahedron
{111} {h k 0} {h l l} where $h>l$ {h h l} where $h>l$	Tetrahedron Pentagonal dodecahedra Tristetra- hedra Deltoid dodecahedra	Octahedron Pentagonal dodecahedra Icositetra- hedra Trisocta- hedra	Tetrahedron Tetrahexa- hedra Tristetra- hedra Deltoid dodecahedra	Octahedron Tetrahexa- hedra Icositetra- hedra Trisocta- hedra	Octahedron Tetrahexa- hedra Icositetra- hedra Trisocta- hedra
General forms {h k l}	Tetrahedral pentagonal dodecahedra	Didodeca- hedra	Hexatetra- hedra	Pentagonal icositetra- hedra	Hexocta- hedra

THE ALLOCATION OF A CRYSTAL TO ITS APPROPRIATE CLASS OF SYMMETRY

We have seen in the course of the above discussion that there is an appropriate general form uniquely characteristic of each of the thirty-two classes of symmetry. The allocation of a particular substance to its proper class by observation of the occurrence of the general form, however, is not always feasible. The crystals may show only special forms, and no special form is uniquely characteristic of one class, though sometimes a particular combination of special forms may be so. Thus in the cubic system a substance occurring only as cubes modified by faces of the rhombic dodecahedron might belong to any one of the five classes in the system, since these are possible special forms throughout the system; but crystals showing a combination of a pentagonal dodecahedron with a tetrahedron must be placed in class 23 even though the development of a tetrahedral pentagonal dodecahedron is never observed (see the Table of Forms above). Even if faces of general forms are present, their precise identification may be a matter of difficulty in systems of low symmetry; to cite an extreme example, how can we tell whether the apparent complete lack of symmetry in the habit of a particular triclinic crystal is due to the presence of pedions only (as in class 1) or to distorted development and suppression of faces, frequent in actual crystals, of some of the faces of pinacoids (class $\bar{1}$)? To meet these difficulties, crystallographers have developed a number of auxiliary lines of approach, of which we shall note four briefly.

1. Etch figures.

Just as the regular geometry and external symmetry of a crystal are an expression of the orderly manner in which the units of construction are built up during the growth of the crystal, so when a crystal is attacked by a suitable solvent the initial dissolution often takes place in a manner which is visibly related to the underlying structure. After contact with the solvent for a period of time (from a few seconds to several hours, depending on its potency and determined by trial) the crystal faces will be seen to show a number of pits or cavities where solution has been most pronounced. The pits are usually bounded by sloping planes along which the solvent has acted most rapidly, and solution cavities of this kind which have definite shapes are known as *etch figures*. Their particular shapes are partly dependent on such factors as the nature of the solvent and its concentration, but the symmetry of their shape and of their attitude on different crystal faces may be considered an indication of the symmetry of the underlying structure. In Fig. 299 are shown diagrammatically a series of etch

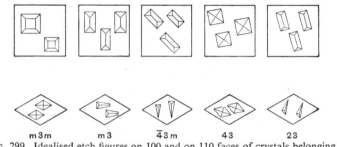

FIG. 299. Idealised etch figures on 100 and on 110 faces of crystals belonging to the various classes of the cubic system.

figures which might be developed on {100} faces and on {110} faces of crystals belonging to each of the five classes of the cubic system. If the etch figures on cube faces, for example, were square in outline but were skew to the outline of the cube face itself, it seems a reasonable inference that the axis normal to the cube face is probably a tetrad axis but that there are no planes of symmetry passing through this axis (class 43). It is important to note, however, that the only safe inference is of the *maximum* probable symmetry of the underlying structure as indicated by any evident lack of symmetry in the etch figures; square pits symmetrically situated on the cube faces might appear on this form in any class of the cubic system, if the solvent develops only etch pits bounded by planes of the rhombic dodecahedron. Etching experiments on cuprite, for example, have always produced etch figures consistent with the full symmetry of class $m3m$, but these results alone cannot be considered a *proof* of holosymmetry, though they point strongly in that direction.

2. Optical activity (optical rotatory power).

The routine examination of the optical properties of a crystal will generally suffice to place the crystal in one of five groups. Cubic crystals are *optically isotropic* for any direction of transmission; crystals belonging to the trigonal, tetragonal and hexagonal systems are *optically uniaxial*, being isotropic only for transmission parallel to the principal axis of symmetry; whilst crystals belonging to the orthorhombic, monoclinic and triclinic systems are *optically biaxial*, and each of these last three systems is distinguished by the extinction-relations in principal zones. Such methods will not afford any indication of the particular class of symmetry, but there is one optical effect which can be shown to be possible only in crystals belonging to certain of the classes of lower symmetry within each system. This effect is known as *optical activity* (or *rotatory polarisation*), and is most easily observed by passing a plane-polarised beam through the crystal parallel to a supposedly isotropic direction—in any direction in a cubic crystal, parallel to the principal symmetry axis of a uniaxial crystal, or parallel to one of the optic axes of a biaxial crystal. If the crystal is optically active it is then observed that the emergent beam, still plane polarised as we should expect, is no longer vibrating parallel to the original direction of vibration, but that the vibration has been rotated during the transmission through the crystal. Different specimens of the same substance may rotate the plane of polarisation in opposite senses, and in many substances (such as tartaric acid, Figs. 183, 184, and quartz, Figs. 218, 219) the sense or ' hand ' of the rotation can be correlated with the morphology of enantiomorphous developments of the crystal. For example, the quartz crystal illustrated in Fig. 218 shows faces of the trigonal trapezohedron $\{5\bar{1}61\}$, developed in zones between the faces of the trigonal bipyramid $\{11\bar{2}1\}$ and the prism $\{10\bar{1}0\}$; if we cut a basal section from such a crystal and transmit through it a beam of plane polarised light we find that the plane of vibration of the emergent beam has been rotated in a right-handed (clockwise) sense as seen by an observer looking towards the source.   The enantiomorphous development shown in Fig. 219 is due to the presence of the trapezohedron $\{6\bar{1}\bar{5}1\}$, in zones between the faces of the bipyramid $\{2\bar{1}\bar{1}1\}$ and of the prism $\{10\bar{1}0\}$; a basal section cut from such a crystal imposes a left-handed (counterclockwise) rotation on the vibration direction of the transmitted plane-polarised beam.   We therefore associate this effect primarily with those classes in which true geometrical enantiomorphism is possible, and for this the class must show no plane of symmetry and no operation of inversion.   The eleven enantiomorphous classes are therefore found in the first row and the sixth row of the chart on p. 106.   Grouped in systems, these classes are:

Triclinic	- - -	Class 1
Monoclinic	- -	„ 2
Orthorhombic	- -	„ 222
Trigonal	- - -	Classes 3, 32
Tetragonal	- -	„ 4, 42
Hexagonal	- -	„ 6, 62
Cubic	- - -	„ 23, 43

Many of the substances which we have quoted above as examples of these classes of symmetry show optical activity, and in some instances this is the chief evidence for allocation to a particular class, since the general form may be rarely, or never, developed.   Thus, in class 42 (p. 132) ethylene diamine sulphate usually crystallises in simple bipyramidal or tabular crystals showing only tetragonal bipyramids and the basal pinacoid and could be holosymmetric, but it is strongly optically active.   It is therefore placed in the enantiomorphous class of the tetragonal system in which tetragonal bipyramids are possible special forms—i.e. in class 42.

Notice particularly that this argument must not be reversed; a substance may possess a structure which results in the symmetry of one of the enantiomorphous classes without this structure being the kind which produces optical activity (see p. 254).   If, therefore, a crystal shows a clear development of an enantiomorphous general form it is correctly allocated to the corresponding enantiomorphous class, even though optically inactive; whilst the lack of optical activity exhibited by cuprite, for example, is no more *conclusive* evidence that its symmetry is not that of class 43 than is the failure to produce asymmetric etch figures.

Finally, we may note that it is theoretically possible * for a crystal

* See, for example, W. A. Wooster, *A Text-Book on Crystal Physics*, Cambridge, 1938, pp. 156–160

belonging to any one of four non-enantiomorphous classes—*m*, *mm*, 4̄, 4̄2*m*—to show optical activity, but no example of this has yet been clearly established.

3. Piezo-electricity and pyro-electricity.

It has been known for a long time that the subjection of certain crystals, such as quartz, tourmaline and hemimorphite (electric calamine), to mechanical stresses causes the separation of electric charges on the surface of the crystal. This phenomenon of *piezo-electricity* (πιέζειν, to press) is now of immense practical importance, but its interest to crystallographers lies primarily in the fact that it is physically possible only in crystals possessing a structure lacking a centre of symmetry. If such a crystal be heated or cooled, electric charges are again developed, the phenomenon being termed *pyro-electricity* (πῦρ, fire), and we can use either of these tests to help in the allocation of a crystal to its proper class—a definite pyro-electric or piezo-electric effect indicates a non-centrosymmetrical class. Thus axinite (p. 109), which is pyro-electric, must be assigned to class 1 in spite of the usual apparent pinacoidal development represented in idealised drawings (Fig. 181), but here again the argument must not be reversed, since it appears that some crystals with a clearly non-centrosymmetrical general form may exhibit no detectable piezo-electric effect.

There are 21 classes without a centre of symmetry:

Triclinic	- - -	Class 1
Monoclinic	- -	Classes 2, *m*
Orthorhombic	- -	„ *mm*, 222
Trigonal	- - -	„ 3, 3*m*, 32
Tetragonal	- -	„ 4, 4̄, 4*mm*, 4̄2*m*, 42
Hexagonal	- -	„ 6, 6̄, 6*mm*, 6̄*m*2, 62
Cubic	- - -	„ 23, 4̄3*m*, 43

but it can be shown that in class 43, as a consequence of the high symmetry, piezo-electricity is not to be expected in spite of the lack of a centre of symmetry.

4. Laue photographs.

For the sake of completeness we describe briefly here the extent to which X-rays can be used to help directly in the determination of the symmetry of a crystal structure. When a beam of X-rays covering a range of wave-lengths is passed through a crystal plate diffraction occurs, as was first demonstrated by M. von Laue in 1912. If the diffracted beams are received on a photographic plate the symmetry of the diffraction pattern (the Laue photograph) thus recorded is related to the symmetry of the crystal structure about the direction of transmission of the original X-ray beam. If this direction were the principal axis of a tetragonal crystal, for example, the photograph would display four-fold symmetry, and if the crystal belonged to class 4*mm* the pattern would also be symmetrical about four planes passing through the tetrad axis. We could thus distinguish a tetragonal class such as class 4 from a ditetragonal class such as class 4*mm*, but the method has an important limitation; it is impossible to tell whether the crystal is centro-symmetrical or not, for the pattern obtained from a non-centrosymmetrical class is indistinguishable from that yielded by a crystal belonging to the higher class generated by the addition of a centre. The thirty-two classes of symmetry give only eleven different groups distinguishable by means of Laue photographs:

Triclinic	-	1 and $\bar{1}$ indistinguishable
Monoclinic	-	2, *m* and 2/*m* indistinguishable
Orthorhombic		*mm*, 222, *mmm* indistinguishable
Trigonal	-	3, $\bar{3}$ show trigonal symmetry
		3*m*, $\bar{3}m$, 32 show ditrigonal symmetry
Tetragonal	-	4, $\bar{4}$, 4/*m* show tetragonal symmetry
		4*mm*, $\bar{4}2m$, 42, 4/*mmm* show ditetragonal symmetry
Hexagonal	-	6, $\bar{6}$, 6/*m* show hexagonal symmetry
		6*mm*, $\bar{6}m2$, 62, 6/*mmm* show dihexagonal symmetry
Cubic	-	23, *m*3 show a two-fold principal axis
		$\bar{4}3m$, 43, *m*3*m* show a four-fold principal axis

THE RELATIVE NUMERICAL IMPORTANCE OF THE VARIOUS SYSTEMS AND CLASSES

It is of interest to enquire how known crystalline substances are distributed over the various systems and classes of symmetry. Precise figures are, of course, impossible of attainment, but the industry of the compilers of works of reference published from time to time enables us to trace the gradual accumulation of accurate data. An early example of such a work, published in 1842, describes about 700 different substances. Groth's *Chemische Krystallographie* (p. 17), published over the period 1906–1919, included about 7,350 substances. At a time when the total had passed 10,000 some estimates were made of the numerical importance of the different systems, and it is interesting to notice that these proportions are preserved, though the actual numbers of substances involved are almost exactly doubled, for the systems so far tabulated in a publication which began to appear in Russia in 1937. It seems that we may place the present total at something approaching 20,000, (of which about 2,000 have been found to occur in nature as minerals and many of these are of extreme rarity). Of this vast number, 50 per cent. belong to the monoclinic system, a further 25 per cent. to the orthorhombic, and about 15 per cent. to the triclinic. Thus these three systems account for about 90 per cent. of the crystal kingdom, leaving only a few per cent in each of the remaining more symmetrical systems. The order of decreasing numerical importance of these seems to be—cubic, tetragonal, trigonal, hexagonal. Within each system, the vastly greater proportion of its representatives belong to the holosymmetric class and, as we have seen above, there is difficulty in finding even a single representative for some of the less symmetrical classes. The numerical unimportance of the more symmetrical systems is a help rather than a hindrance so far as determinative problems are concerned, for whilst all cubic crystals of the same symmetry have identical interfacial angles the number of variables increases steadily as the symmetry diminishes.

CHAPTER VII

PARALLEL GROWTH AND COMPOSITE CRYSTALS

So far, discussion has centred almost entirely on individual crystals. In practice one is often concerned with *crystalline aggregates*; in a crystallisation taking place, for example, in an open vessel the first crystals to form may adhere to the walls or float freely in the solution, but, if much further material separates, an aggregate of more or less interlocking crystals is ultimately produced. Adjacent crystals in such an aggregate may be in quite haphazard juxtaposition, but frequently one observes a strong tendency to parallel growth, manifested by the parallelism of corresponding edges and corresponding faces of various individuals. Octahedra of alum, for example, frequently have the arrangement of Fig. 300, a number of individuals being associated with

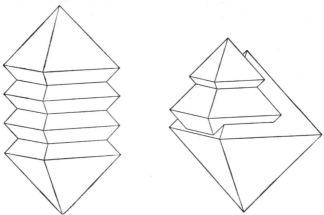

Fig. 300-301. Parallel growths of octahedra.

one tetrad axis in common and the other crystallographic axes parallel throughout the group. A less symmetrical parallel grouping of octahedra is represented in Fig. 301, where individuals of different sizes are aggregated with the tetrad axes in parallel position throughout.

Crystals of cubic habit are often grouped together in branching (*dendritic*) aggregates extending in the directions of the four triad axes; such a dendritic group, growing outwards from the centre (Fig. 302), may gradually fill up during crystallisation to give a single large cube,

but the manner of growth is frequently revealed by slight irregularities
in the placing of the individuals, and these irregularities, resulting in a
departure of the cube-faces of the large composite individual from a

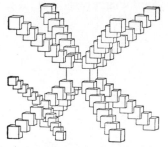

FIG. 302. Parallel growth of cubes
along triad axes.

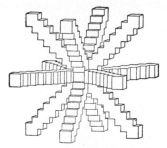

FIG. 303. Parallel growth of cubes
along diad axes.

truly plane surface, remain to reveal the *lineage-structure* of the whole
composite growth.   Variations of this in detail are found where the
aggregation has been along the directions of the diad axes (Fig. 303).
In still other cases, the cubelets may be grouped to simulate a large
octahedron (Fig. 304).   Where the particular conditions of growth tend

FIG. 304. Cubes in parallel growth
simulating an octahedron.

FIG. 305. Parallel growth formed during
crystallisation on a plane surface.

to emphasise a particular plane, as for example when a shallow drop
of solution is evaporating on the surface of a micro-slide, the parallel
growth may extend along certain directions in this plane; the crystals
of Fig. 305 are all orientated with a triad axis normal to the plane of
the figure and are aggregated along the direction of the three diad axes
in this plane.   In parallel growths of this kind, individual crystals are
frequently very distorted dimensionally, and some of the faces of the
different forms present are altogether suppressed; in the aggregate of

copper crystals in Fig. 306 the directions of elongation of individuals, as also of aggregation throughout the growth, are all parallel to cube-dodecahedron or cube-octahedron edges (see inset on figure). A similar dimensional distortion with occasional suppression of faces has

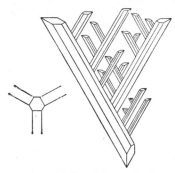

FIG. 306. Parallel growth of copper crystals.

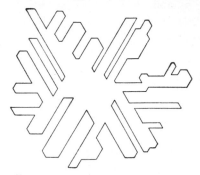

FIG. 307. Parallel growth of hematite.

produced the composite platy crystal of Fe_2O_3 (hematite) illustrated in Fig. 307 by a plan on 0001. A familiar example of this mode of aggregation in hexagonal crystals is found in the myriad variety of grouping of ice-crystals in ' snow crystals ' (Fig. 308) and arborescent frost-growths.

Under some conditions of growth there may be a tendency for material to be added more rapidly at some points on the face of a growing crystal than at other points on the same face. A crystal lying with a face in contact with the bottom of the vessel containing the saturated solution, for instance, easily receives further material around

FIG. 308. A ' snow crystal.'

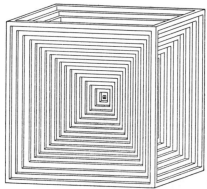

FIG. 309. An idealised hopper cube.

the edges of that face by diffusion of the saturated solution towards the growing crystal, but if crystallisation is rapid little or none of this solution may penetrate to the central area of the face. The crystal grows by addition of layers of material around the edges, rather than by addition of layers extending right across the face in question, and the completed crystal is in part skeletal. An idealised drawing of such a skeletal cube (a ' *hopper crystal* ') is shown in Fig. 309; in practice one may frequently find that certain faces of a form have a more pro-

FIG. 310. Hopper development of a rhombic dodecahedron.

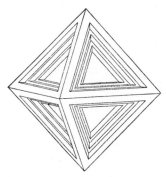

FIG. 311. Hopper development of an octahedron.

nounced skeletal development than others. Fig. 310 shows a similar habit in the rhombic dodecahedron. In the octahedron of Fig. 311 the hopper development is due to the appearance of a second form, the steps being built up by an alternation of octahedral and cubic

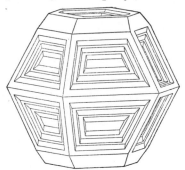

FIG. 312. A crystal of sulphur showing partial hopper development.

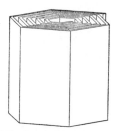

FIG. 313. A spiral hexagonal crystal, such as ice sometimes builds.

planes. In the crystal of sulphur illustrated in Fig. 312 the faces of one bipyramid show no tendency to this skeletal development. The student may eventually encounter many varieties of this type of skeletal

P.C.   L

growth amongst crystalline substances; one further example is portrayed in Fig. 313, where a hexagonal crystal has grown as a hollow hexagonal prismatic spiral.

Alternating development of the faces of different forms sometimes produces the appearance exemplified by the octahedron of Fig. 314. The octahedral habit is built up by the addition in each octant of layers parallel to the octahedral planes, but bounded by planes of the rhombic

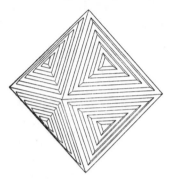

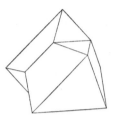

Fɪɢ. 314. Alternating development of octahedron and rhombic dodecahedron.

Fɪɢ. 315. A twinned octahedron.

dodecahedron. Crystals with regularly striated faces often result from this kind of alternating development; with differing relative developments of the two forms present, Fig. 314 might present the appearance of an octahedron with faces bearing equilateral triangular striations or of a rhombic dodecahedron with faces striated parallel to the longer diagonal of each rhombus.

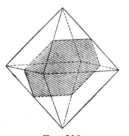

A different kind of composite crystal from those which we have so far considered is represented in Fig. 315. The crystal is clearly related to an octahedron, but the front lower portion is in a reversed position. If we imagine an octahedron divided symmetrically by a plane ȷarallel to one pair of external faces (Fig. 316), the crystal of Fig. 315 could be obtained by rotation of the lower portion through 180° about

Fɪɢ. 316.

the triad axis normal to the face 11$\bar{1}$. A crystal of this kind is called a *twin crystal* (German *Zwilling*, French *macle*). Each portion of a twin crystal has its own proper orientation, but the two orientations are simply related to the crystallography of each portion. Usually such a composite crystal has grown as a twin crystal, with the two orientations present from the beginning of crystallisation, but it is convenient in description to talk of the operation of ' rotation ' which would bring one orientation into congruence with the other.   (In some substances, a related kind of composite crystal, usually also called a twin crystal, can be produced by mechanical deformation after growth, but such *secondary twins* show important differences from the growth twins under discussion here.)   The line about which the rotation, necessary to achieve congruence, must be made is called the *twin-axis*, and in the most frequent type of twinning it is normal to an important crystal face.   *Re-entrant angles*, such as those of Fig. 315, are frequently an indication of twinning, simple crystals usually presenting only *salient* interfacial angles.

The differing orientation of the two portions of the twinned octa-hedron of Fig. 315 could also be related by reflection in the shaded plane of Fig. 316, parallel to the face 11$\bar{1}$, and a plane related in this way to the two portions is called a *twin-plane*.   Usually the concept of reflection in a twin-plane affords the easiest picture of the morphology of the twin, whilst we shall find later that the concept of rotation about a twin-axis is more readily applied to the study of stereograms of twin crystals.   Where the simple crystal belongs to a centro-symmetrical class of symmetry the operation of rotation through 180° about a twin-axis and the operation of reflection across a twin-plane normal to this axis produce identical results, and either description can be used at will —the crystal of Fig. 315 may be described as twinned about a triad axis or, alternatively, as twinned on a face of the octahedron.   In crystals lacking a centre of symmetry, however, these operations do not produce identical results, and care must be taken to use the description which correctly applies to the twin in question.   The crystal shown in Fig. 317 is composed of the two complementary tetrahedra {111} and {1$\bar{1}$1}, the faces of the latter form being shaded to emphasise the structural difference between the planes 111 and 1$\bar{1}$1.   The twin obtained

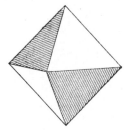

FIG. 317. A combination of the tetrahedra {111} and {1$\bar{1}$1}.

by rotation about the triad axis normal to the face 11$\bar{1}$ is illustrated in Fig. 318, and this is clearly not identical with a reflection-twin on 11$\bar{1}$ (Fig. 319).   The reflection-twin is described as a *symmetric twin*, whilst for the rotation-twin the obsolescent term *hemitrope*, introduced by Haüy, may still be used ($\acute{\eta}\mu\iota$ = half, $\tau\rho\acute{o}\pi os$ = a turn).

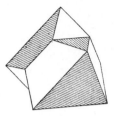

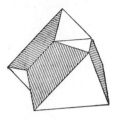

Fig. 318. A rotation twin of the crystal of Fig. 317.

Fig. 319. A reflection twin of the crystal of Fig. 317.

The plane along which the two portions of a twin crystal appear to be united is called the *composition-plane*. (There is usually no physical discontinuity marking this plane, the change across it being merely one of change of orientation of the structure; sometimes, however, it is a plane of easy splitting—a *parting*.) In Fig. 315 the composition-plane is identical with the twin-plane, and this is frequently the case, but not always. The twin crystal illustrated in Fig. 320 clearly consists of an upper and a lower portion united on a composition-plane parallel to the basal pinacoid, but the disposition of the faces of the shaded form reveals that this plane is not the twin-plane. The twin-plane

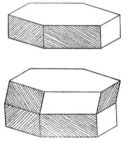

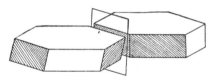

Fig. 320. A twin of the mono-clinic (pseudohexagonal) crystal illustrated in the upper portion of the figure. The shaded faces belong to the form {110}.

Fig. 321. The reflection operation which gives the two orientations shown in the twin crystal in Fig. 320.

must be normal to the basal pinacoid (Fig. 321); or, in other words, the twin-axis lies *in* the composition-plane. This kind of twin is called a *parallel twin*, in contrast to the *normal twins* which we have described above. Parallel twins are much less frequently found than normal twins; for the sake of completeness, we must mention briefly a third kind of twin, still more rarely found, in which the twin-axis is not a possible crystal edge, but is a line in a crystal plane normal to a possible edge.

The three types of rotation-twin are thus:

Normal twins.   The twin-axis is the normal to a possible crystal face, the twin plane, which is the composition-plane of the twin.

Parallel twins.   The twin-axis is a possible crystal edge (a zone axis) lying in the composition-plane (which is not necessarily a possible crystal face).

Complex twins.   The twin-axis is a line in the composition-plane normal to a possible crystal edge.

It will be seen from the foregoing discussion that an axis of symmetry of even degree (2, 4 or 6) cannot function as a twin-axis, since no new orientation is produced by rotation through 180° about such an axis. Nor can a plane of crystallographic symmetry function as the twin-plane in normal twinning, though it may be the composition-plane in parallel or in complex twinning.

In the twins so far described the two orientations have developed one on either side of a well-defined plane; the impression received is of the two portions in contact along this plane, and such twins are described as *contact-twins*. The tetrahedra making up the composite crystal illustrated in Fig. 322 show no such plane of contact, but rather appear to be grown through each other; the crystal is an *interpenetrant-twin*. The concept of a composition-plane is here no longer applicable, but the two orientations are still related by the appropriate operation of rotation or of reflection.

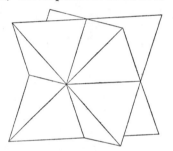

FIG. 322. An interpenetrant twin of tetrahedra.

Twin crystals of sodium chlorate, class 23, with the habit of Fig. 322 are deposited from aqueous solution at low temperatures, and the two tetrahedra are found to be built on structures showing opposite hands of optical activity; the two orientations must therefore be related by an operation of reflection, and the twin is correctly described as a reflection-twin on a face of the cube {100}—it is a symmetric twin and not a hemitrope. The distinction of contact-twins from interpenetrant-twins on the score of their outward appearance is a convenient one in practice, but there is a complete transition possible from one to the other. Examination of thin sections of twinned crystals reveals continuous gradations from types in which the two orientations are developed on either side of a clearly-defined plane, through examples in which the dividing surface is slightly irregular, to completely interpenetrant-twins in which the two orientations are intimately intergrown in the central portions. Yet, however intimate the intergrowth, the orientation of the structure at any particular point is always that of one or the other of the two attitudes in question.

The reasons for the formation of twin-crystals, as for many other features of crystal habit, are not yet completely understood. Some substances are never found as twins, whilst others may scarcely ever crystallise as simple individuals. Others, again, are found sometimes twinned but sometimes simple, and it is clear that the environment during crystallisation plays some part. The most important factor, however, is believed to lie in the geometry of the internal structure of the particular substance, the substance being by chance almost symmetrical about the direction of a twin-axis (*axis of pseudo-symmetry*) or almost symmetrical over a twin-plane (*plane of pseudo-symmetry*). In the structural pattern a unit can be marked out, larger perhaps than the true smallest unit of pattern, which approximates in symmetry to a higher class; when a portion of the structure is in twinned orientation this pseudo-unit extends almost without deviation over the two orientations alike. Though we are not yet in a position to examine this proposition in relationship to the details of the internal structural pattern, we shall often find that the pseudo-symmetry is already clearly suggested by the external morphology of the simple crystal.

Fig. 323 represents a simple crystal of potassium sulphate, K_2SO_4, class *mmm*. The angle $110 \wedge 1\bar{1}0 = 59° 36'$, so that $110 \wedge 010 = 60° 12'$,

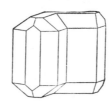
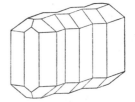

FIG. 323. A crystal of potassium sulphate.

FIG. 324. A contact twin of potassium sulphate.

FIG. 325. Repeated contact twinning.

and the prism-zone is thus morphologically pseudo-hexagonal; in agreement with the above suggestion, potassium sulphate shows a strong tendency to twinning on {110} planes. Fig. 324 is a simple contact twin of this kind, whilst in the crystal illustrated in Fig. 325 the twinning is repeated. In Fig. 326 the twinned individuals interpenetrate, and where three different orientations are present we may also find either a contact twin (Fig. 327) or an interpenetrant association (Fig. 328). To emphasise the number of individuals composing such twins

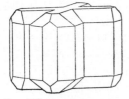
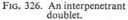
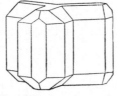
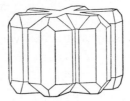

FIG. 326. An interpenetrant doublet.

FIG. 327. A contact twin of three individuals.

FIG. 328. An interpenetrant triplet.

we may speak of *doublets, triplets*, etc., but repeated change of orientation is generally termed, rather loosely, 'multiple twinning'. The plan on the basal plane of an interpenetrant triplet (Fig. 329) shows how such a group may be described as an association, in twinned orientation, of individuals II and III on either side of individual I. In such

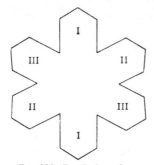
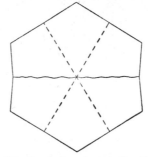

FIG. 329. Basal plan of an interpenetrant triplet.

FIG. 330. Basal plan of a twin like Fig. 329 with the deep re-entrant angles filled up.

twins, there is a strong tendency for the re-entrant angles between adjacent {010} faces to fill up during crystallisation. The {110} faces of individual I are co-planar with faces of the same form in individuals II and III, so that the only external indication of twinning in a completely filled-up crystal is afforded by the shallow re-entrant angles on the two side faces (Fig. 330). In substances which are closely pseudo-hexagonal in simple crystals, these re-entrant angles amount only to a few minutes of arc, and a twinned crystal imitating so nearly the symmetry of a higher system is called a *mimetic twin* (Fig. 331).

Imitative twinning of this kind is very frequent in orthorhombic pseudo-hexagonal substances, and we can discuss some further possibilities most easily in terms of plans on the basal plane. The direction of the x-axis (the zone-axis of the zone between 010 and 001) is frequently marked by striations in this direction on the basal pinacoid

(Fig. 332). In a regular interpenetrant triplet such as Fig. 330 the nature of the twinning is then rendered evident by the arrangement of V-shaped striations (Fig. 333), but this feature often serves to show

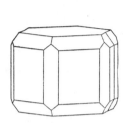

FIG. 331. Clinographic view of a mimetic twin of potassium sulphate.

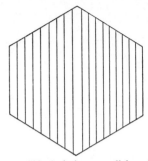

FIG. 332. Striations parallel to the x-axis on the 001 plane of an ortho-rhombic pseudohexagonal crystal.

FIG. 333. V-shaped striations on the basal plane of an interpenetrant triplet.

FIG. 334. A twin like Fig. 333, less regularly developed.

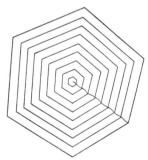

FIG. 335. As a result of repeated contact twinning the outline of this twin is formed entirely by 010 faces.

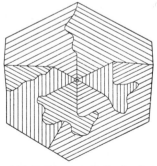

FIG. 336. In this example the development is quite irregular, and the directions of striations show that the bounding planes belong partly to {010} and partly to {110}.

that the association of the different orientations is much more irregular (Fig. 334). In the ideal twin of Fig. 333 the prism faces of the twin all belong to the form {110} of the variously orientated individuals, but a variation of this arrangement may lead to a mimetic twin bounded entirely by faces of {010} (Fig. 335) or by faces belonging partly to {110} and partly to {010} (Fig. 336).

STEREOGRAPHIC PROJECTION OF TWIN CRYSTALS

By the 'operation' of twinning, the spherical projection (from which the stereographic projection is derived) is rotated through 180° about the twin-axis. By this rotation, every great circle passing through the twin-axis is turned around to coincide with itself in reversed position. On the stereogram, the traces of all such great circles pass through the pole of the twin-axis. Thus, to obtain the twinned position of a pole P (Fig. 337), draw the great circle through P and the pole of

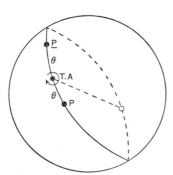

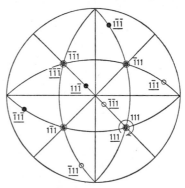

FIG. 337. Construction of the twinned position $\underline{P}$ of a pole P at an angular distance θ from the pole of the twin axis T.A.

FIG. 338. Stereogram of an octahedron twinned about the normal to 111. Only the four upper faces of the original orientation are indexed, but the position of all eight faces in the twinned orientation are shown by underlined indices.

the twin-axis, and mark the pole $\underline{P}$ in the same great circle at an equal angular distance on the opposite side of the twin-axis. Then $\underline{P}$ is the twinned position of P.

The stereogram in Fig. 338 shows in this way the twinned positions of the poles of an octahedron twinned about the triad axis normal to the plane 111. The pole 1$\bar{1}$1, for example, is at an angular distance of 70° 32' from the pole of the twin-axis, and is at this same angular distance on the opposite side after twinning, in the great circle through 1$\bar{1}$1–111–010. Since 111$\wedge$010 = 54° 44', the twinned position of 1$\bar{1}$1 is on the lower hemisphere at an angular distance 70° 32′–54° 44′ from 010, in the position marked 1$\bar{1}$1. This convention of underlining the index of a pole in a twinned position is widely followed. Before the relationship of twinning to a pseudo-symmetry of the underlying structure was clearly understood, it was believed that faces in a twinned position could be given new rational indices related to the original set of crystallographic axes. It is now realised that such a supposition is groundless, and that the faces of the portion of a crystal in twinned orientation must be given indices related to its own set of crystallographic axes, derivable by the operation of twinning from the original set. A symbol such as 1$\bar{1}$1, therefore, must be read ' the position to which the face 1$\bar{1}$1 of the original crystal is brought by the operation of twinning adopted to describe the crystal under consideration '.

   Though the general construction described above will always suffice to project the twinned position of a pole, we can sometimes proceed more easily by first constructing the twinned position of an arc on which the required pole must lie.   Fig. 339 is a projection of an orthorhombic crystal in which the zone of domes $\{0\,k\,l\}$ is pseudo-hexagonal.   In chrysoberyl, $BeAl_2O_4$, for example, $0\bar{1}1 \frown 011 = 60° 14′$, and we might confidently expect twinning on a plane of the form $\{011\}$ to yield a mimetic twin.   Such twinning would bring faces of the bipyramid $\{111\}$ into strict parallelism in adjacent portions of the twin—$\bar{1}$11, for example, twinned about the normal to 0$\bar{1}$1 would coincide with 1$\bar{1}$1. Examination of actual twins reveals that there is a minute re-entrant angle between adjacent bipyramidal faces, and we must seek a different explanation of the twinning.   Since $001 \frown 011 = 30° 7′$, therefore $001 \frown 031 = 60° 7′$, and the pole 031 is 90° 14′ from the pole 0$\bar{1}$1, 89° 46′ from the pole 01$\bar{1}$.   A rotation cf 180° about the normal to 031 will therefore bring 01$\bar{1}$ almost into coincidence with 0$\bar{1}$1, but not quite (the pole 01$\bar{1}$ is displaced outwards in the figure for clarity). $\bar{1}$00 coincides with $\overline{100}$, so that the twinned position of the zone 100–0$\bar{1}$1–$\bar{1}$00 almost coincides with the original zone 100–0$\bar{1}$1–$\bar{1}$00.   The pole $\bar{1}$1$\bar{1}$ lies on this zone at an angular

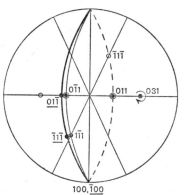

FIG. 339. Stereogram to illustrate twinning about the normal to a plane $0\,k\,l$ in an orthorhombic crystal.

distance from 100 equal to the angle $100 \wedge 1\bar{1}1$; it therefore makes a minute angle (19′ in chrysoberyl) with the pole $1\bar{1}1$, and the existence of this re-entrant in the actual specimens is adequately accounted for by a description of the twinning in terms of the normal to the face 031 as twin-axis.

COMMON TYPES OF TWIN IN THE DIFFERENT CRYSTAL SYSTEMS

Cubic system.

The crystallographic axes (normals to cube faces) are symmetry axes of even degree in every class of the cubic system, and thus cannot function as twin-axes according to the description of twinning which we have presented here. (A face of the form {100} may be a twin-plane of the rare symmetric twins in those classes lacking the cubic planes of symmetry, as described above (p. 165) for sodium chlorate.) In class $m3m$, the normals to the faces of {110} are diad axes, and the triad axes are the first important crystallographic directions likely to function as twin-axes. Cubes twinned about a triad axis may be associated as contact-twins (Fig. 340), as in copper, or may be interpenetrant, as in

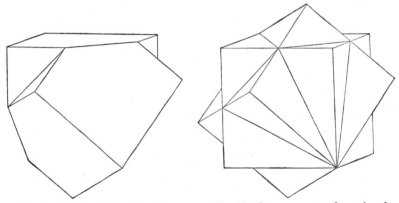

FIG. 340. Contact twin of a cube twinned about a triad axis.

FIG. 341. Interpenetrant cubes twinned about a triad axis.

fluorspar, CaF_2 (Fig. 341). Similar variations are presented by crystals of dodecahedral habit twinned on the same law (Figs. 342, 343). The twinned octahedron, which we have already discussed (Fig. 315), is usually a contact twin; it is very characteristic of a group of substances $R''R_2'''O_4$ named, from the member $MgAl_2O_4$, the Spinels, and hence is usually called the spinel twin. It is found in many

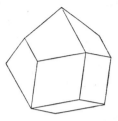

FIG. 342. Contact twin of rhombic dodecahedra twinned about a triad axis.

FIG. 343. Interpenetrant dodecahedra twinned about a triad axis.

other cubic substances, such as gold and diamond, the twin crystals being often tabular in habit as a consequence of extension of the crystal faces parallel to the composition-plane (Fig. 344). The twinning may be repeated, either about the same triad axis (Fig. 345) or about different triad axes (Fig. 346). In classes 23, $m3$ and $\bar{4}3m$, in which

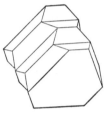

FIG. 344. A spinel twin flattened parallel to the composition plane.

FIGS. 345-346. Repeated spinel twinning.

the normal to a dodecahedral face is not a diad axis, this direction may function as a twin-axis, as for example in interpenetrant tetrahedra (Fig. 322) in the non-enantiomorphous class $\bar{4}3m$. Occasionally other twin-laws are found, as on 211 planes (sodium uranyl acetate) or 441 planes (galena).

Tetragonal system.

The common twinning in crystals of the holosymmetric class, $4/mmm$ has a face of a tetragonal bipyramid, $\{h\,0\,l\}$ or $\{h\,h\,l\}$, as twin-plane, often a face of the form indexed $\{101\}$. A simple contact twin of a prismatic crystal on this law produces a *geniculated crystal* or *elbow-twin* (Fig. 347). Repeated twinning may take place in such a manner that the z-axes of the various portions in different orientations all lie in one plane (Fig. 348). It may also occur partly on planes of type 011 and partly on planes 101 of the same form; the z-axes of the

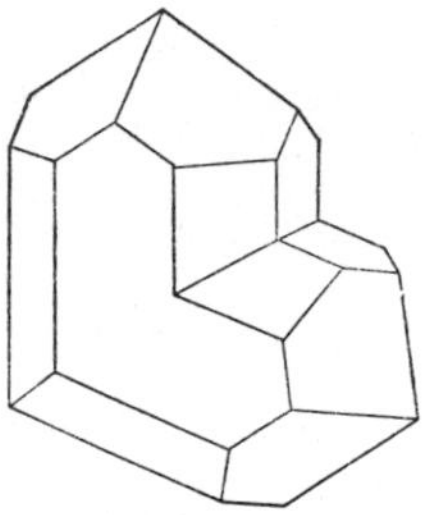

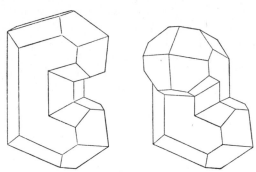

FIG. 347. Elbow twin of a tetragonal crystal.

FIGS. 348-349. Repeated twinning of the type shown in Fig. 347.

portions of a multiple twin of this kind (Fig. 349) do not all lie in the same plane, and in this way repeated twinning sometimes gives rise to remarkable composite crystals, as in the minerals rutile, TiO_2, and cassiterite, SnO_2. Twins on this law may also be interpenetrant, as in calomel, HgCl (Fig. 350). If the axial ratio c/a of a tetragonal crystal

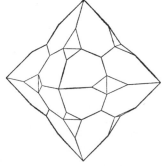

FIG. 350. A twin crystal of mercurous chloride.

FIG. 351. Pseudo-octahedral aspect of a twinned tetragonal crystal.

is near unity, the crystal is morphologically pseudo-cubic and we may expect mimetic twinning. By twinning on 011 the tetrad axis is turned almost through 90°, and an interpenetrant triplet may in such a case simulate very closely a cubic habit (Fig. 351).

In classes of lower symmetry, further possibilities arise. We have quoted iodosuccinimide above as an example of a substance belonging to class 4, but it also crystallises with the habit of Fig. 352, in which the tetrad axis is apparently not uniterminal. Such crystals are twinned by reflection in the basal pedion, and illustrate once again how twinning can be regarded as an apparent attempt to achieve a higher symmetry; for the basal pedion is not a plane of symmetry of the simple crystal, whilst the twin does apparently possess a symmetry plane in this direction. If re-entrant angles are visible, as in the figure, the twinning is easily detected, but if these are filled up during growth the crystal would appear externally as a simple individual of bi-pyramidal habit. (Mimetic twinning of this kind, which raises the apparent symmetry of a crystal to a higher class within the same system, is sometimes distinguished as *supplementary twinning*.) The presence of twinning in such a crystal would be revealed by etch-figures, which might be asymmetric pits pointing in opposite directions on the upper and lower portions of an apparently single prism-face; or by a test for pyro-electricity, the upper and lower terminations showing similar electrification with a region of opposite sign around the middle of the crystal.

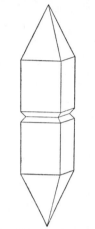

Fig. 352. A twin crystal of iodosuccinimide.

Orthorhombic system.

We have already described the frequent twinning on a prism-face (or a dome, according to the setting of the crystal) in substances in which the prism-zone is morphologically pseudo-hexagonal. If the angle $1\bar{1}0 \wedge 110$ of an orthorhombic substance is approximately 90°, conferring a pseudo-tetragonal habit, we may likewise find mimetic twinning on a face of the form {110} to increase the resemblance to a tetragonal substance. In the minerals of the marcasite group the prism-angle is near 72°, and it is interesting to find repeated twinning on a prism-face is frequent here also, for the pseudo-pentagonal 'fivelings' thus produced do not, of course, simulate a possible higher crystallographic symmetry.

Twinning on faces of an orthorhombic bipyramid, {h k l}, is not common, though sometimes found in association with twinning on a face of a dome or prism; chalcocite, Cu_2S, for example, shows twins on 110 (mimetic, pseudo-hexagonal), on 032 and on 112; staurolite on 032, on 230 and on 232.

In the hemimorphic class *mm*, supplementary twinning on a basal pedion may raise the apparent symmetry to that of class *mmm*, and is described in many of the substances which we have used (p. 113) as illustrations of this class.   Fig. 353 depicts a twin of struvite, and Fig. 354 one of bismuth thiocyanate, on this law.   If repeated twinning of this kind were present on a very fine (perhaps submicroscopic) scale in a crystal of class *mm*, it would be difficult to determine readily

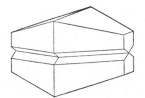

FIG. 353   Supplementary twin of struvite.

FIG. 354.   Supplementary twin of bismuth thiocyanate.

that it lacked the higher symmetry of class *mmm*.   The silicate topaz $(OH,F)_2Al_2SiO_4$, apparently affords an example of this, since, although many specimens show a bipyramidal habit and yield symmetrical etch-figures, others give a definite pyro-electric and piezo-electric effect, especially if crushed into fragments before testing; presumably, there-fore, the true unit of the underlying structure has the symmetry of class *mm* only, but many natural crystals are built up by intimate supplementary twinning.

Twinning in crystals of class 222 is rare, though symmetric twins of enantiomorphs, related by reflection in a face of the prism {110}, have been described in some substances.

Monoclinic System

The only crystallographic line which cannot function as a twin-axis in this system is the diad axis of classes 2 and $2/m$, and the plane of symmetry is likewise inadmissible as a twin-plane in classes *m* and $2/m$. The commonest twins are those on planes parallel to the *y*-axis (planes $h\,0\,l$, including those indexed 100 and 001) or about a crystal edge normal to the *y*-axis.   The crystal of potassium chlorate, $KClO_3$, shown in Fig. 355, is a simple contact twin on 001; this twinning is often repeated, so that an apparently simple crystal is found to be built up from a series of thin lamellae alternating in orientation (Fig. 356). Such repeated twinning is described as *polysynthetic*, and seems likely to be developed most finely in substances in which the pseudo-structural unit (p. 166) extends across the composition-plane with the least possible deviation. In extreme cases, the individual lamellae may be

FIG. 355. Contact twin of potassium chlorate.

FIG. 356. Polysynthetic twinning in potassium chlorate.

of submicroscopic width, and the crystal may acquire a plane of apparent symmetry parallel to the composition-plane—a further variety of mimetic twinning. (Secondary twinning (p. 163) is often polysynthetic.) Gypsum, $CaSO_4.2H_2O$, may be quoted as an example of twinning on more than one face in the zone parallel to the y-axis. Twins on 100 are common, either as contact-twins (Fig. 357) or interpenetrant (Fig. 358), but rather similar ' swallow-tail ' twins also occur in which the twinning is on the face 101 (Fig. 359).

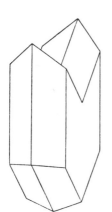

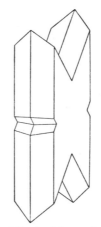

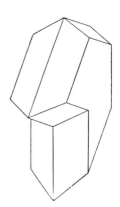

FIG. 357. A contact twin of gypsum, twinned on 100.

FIG. 358. An interpenetrant twin of gypsum; the twin law is the same as in the crystal in Fig. 357.

FIG. 359. A contact twin of gypsum, twinned on 101.

Mica affords an example of parallel twinning in a monoclinic crystal, with composition-plane 001 (Fig. 320); the Carlsbad twin in orthoclase is also a parallel twin, the twin-axis being the z crystallographic axis and 010 the composition-plane.

In classes 2 and *m*, supplementary twinning on the plane 010 or about the *y*-axis respectively may raise the apparent symmetry to that of class 2/*m*.

Triclinic System

Since no planes of symmetry and no axes of symmetry are present in this system, twinning can occur on any crystal plane or about any crystallographic direction. Usually a twin-plane will be an important crystallographic plane in the most convenient setting of the crystal, such as faces of the pinacoids {100}, {010} or {001}. Polysynthetic twinning may occur on two or more such planes, and is often mimetic. Such twinning has been studied in great detail in some minerals; the plagioclase felspars, for example, show at least twelve types of twin on normal, parallel and complex laws. Supplementary twinning may account in some instances for the apparent pinacoidal development of crystals which prove to be pyro-electric, and which must therefore be allocated to class 1.

Hexagonal System

Since there is no crystal class with more than one hexad axis, it is perhaps theoretically unlikely that twinning will often occur in such a way as to produce a composite group with inclined hexad axes. The hexad axis itself, being an axis of even degree, cannot function as a twin-axis; nor can the plane normal to it be a twin-plane in the classes in which there is a horizontal plane of symmetry. In agreement with these suggestions, the only important twinning in the hexagonal system is supplementary twinning on the plane 0001 in the classes 6 and 6*mm*, in which the hexad axis is uniterminal. The true symmetry of such a twin is revealed, as in the corresponding cases in the tetragonal system, by the disposition of pyro-electric charges or by the nature of the etch-figures. Fig. 360 illustrates the appearance sometimes presented by crystals of zinc oxide, ZnO, in which two pyramidal crystals of different sizes are united in twinned orientation.

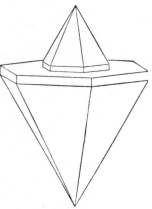

Fig. 360. Crystals of zincite in twinned orientation.

Trigonal System

The chief interest of twins in this system lies in the fact that the principal axis of symmetry, the triad axis, can function as a twin-axis since

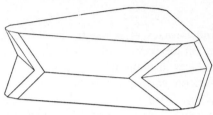

it is an axis of odd degree. Thus even in the holosymmetric class $\bar{3}m$ a variety of twins are found corresponding to hemitropy about the triad axis (or twinning over the plane 0001). Fig. 361 illustrates a contact-twin of this kind in chromium oxide, Cr_2O_3, of tabular habit;

FIG. 361. A twin crystal of chromium oxide.

Fig. 362 is a twin on the same law of a scalenohedral crystal of calcite, $CaCO_3$. An interpenetrant development is frequently found in crystals of rhombohedral habit (Fig. 363). In the enantiomorphous classes, careful distinction must be made between twins by rotation and twins by reflection. Quartz, for example, in class 32, shows different kinds

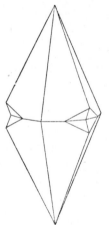

FIG. 362. A twinned scalenohedron.

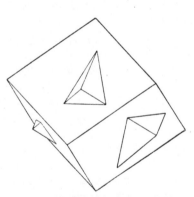

FIG. 363. Interpenetrant twin of rhombohedra.

of twin in which the triad axis has the same direction for both orientations; in the Dauphiné twin, parts in different orientation are of the same hand, and the twinning is correctly described as hemitropy about the triad axis, whilst in the Brazil twin structures of opposite hand are found in twin-association—the twin is a symmetric twin by reflection in a face of the prism $\{11\bar{2}0\}$. Both types are usually irregularly interpenetrant.

Fig. 364 is an example of parallel twinning in a crystal of $Na_3Li(SO_4)_2 . 6H_2O$, class $3m$. Though the composition-plane is 0001,

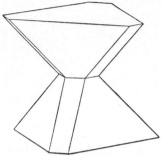

the two portions are clearly not in orientations related by reflection in this plane; the twin-axis is the normal to a face of the prism $\{11\bar{2}0\}$, and is thus a possible crystal edge in the plane 0001. Twinning on this law is shown also by pyrargyrite and proustite; if the two portions are grown together without re-entrant angles, the hemimorphic character of the true symmetry ceases to be evident and the apparent symmetry is raised to that of class $\bar{3}m$. In

FIG. 364. Parallel twinning in a trigonal crystal.

a few substances in which this twin law is known, normal twinning on the plane 0001 has also been observed (Fig. 365).

Twinning on a face of a rhombohedron produces a twin-association in which the triad axes of the two orientations are inclined at an angle depending upon the slope of the twin-plane. In calcite,

$$0001 \wedge 10\bar{1}1 = 44° \, 36\tfrac{1}{2}',$$

and the triad axes are almost at right-angles (Fig. 366). In such twins, there is frequently a tendency for the re-entrant angles between the two portions to fill up during crystallisation by the extension of certain faces of a form at the expense of others (Fig. 367); the continuation

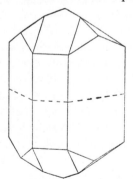

FIG. 365. Mimetic twin of a trigonal crystal.

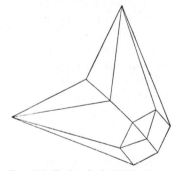

FIG. 366. Scalenohedra twinned on a rhombohedral plane.

of this process may lead to the complete suppression of some faces, and the interpretation of the morphology of the resultant 'butterfly twin' (Fig. 368) is not always evident at first sight.

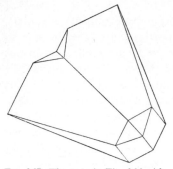

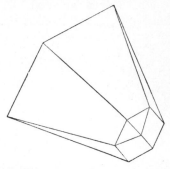

FIG. 367. The twin in Fig. 366 with the re-entrant angle partly filled up.

FIG. 368. A ' butterfly ' twin, in which the re-entrant angle is completely filled.

As in other systems, supplementary twinning of substances built on a non-centrosymmetrical structure may explain the observation of anomalous pyro-electric effects in crystals (such as dioptase) with apparently centrosymmetrical morphology.

THE FREQUENCY OF OCCURRENCE OF TWINNING

We have already pointed out that, whilst some substances have never been observed in twinned aggregates, twinning is so constantly present in others that this feature is of distinct diagnostic value, and we have suggested that the frequency of twinning is related to the degree of pseudo-symmetry exhibited by the underlying structure. It is of interest to enquire how widespread such pseudo-symmetry may be. A review of the whole crystal kingdom is a task of enormous magnitude, but mineralogists have paid close attention to the study of twinning in naturally-occurring minerals, and we confine our observations to this field. Twinned associations have been described in about 20 per cent. of known minerals; two-thirds of these twinned minerals show twinning on one law only, whilst in the remainder two or more laws are known. Normal twins are ten times more frequent than other types. Amongst the different systems, twinning is most important in the monoclinic and orthorhombic, and declines in importance through the cubic, trigonal, tetragonal, triclinic and hexagonal systems. We may remark again that, as with other features of crystal habit, the environment during crystallisation plays a certain part in determining the abundance of twinned individuals in a particular crop.

CHAPTER VIII

SOME MATHEMATICAL RELATIONSHIPS

One reason for the extensive use of the stereogram for the representation of crystals lies in the fact that since all circles drawn on the sphere are represented as circular arcs in stereographic projection, the stereogram lends itself readily to graphical work. The accuracy of the measurements which we can carry out with the reflecting goniometer, however, justifies an order of accuracy in subsequent computations higher than that which we can easily maintain in graphical work; an equally important feature of the stereogram is the ease with which we can carry out calculations upon it by using the formulae of spherical trigonometry. A *spherical triangle* is a figure on the surface of a sphere (in crystallography the spherical projection) bounded by the arcs of three great circles. Thus all the triangles into which a stereogram is divided by zone-circles are representations in projection of spherical triangles. The arcs are termed *sides* of the triangle; the three sides and three angles collectively make up the *parts* (or *elements*) of the triangle. In Fig. 369, the arcs *a*, *b*, *c* are the sides and the angles *A*, *B*, *C* the angles of the spherical triangle *ABC* inscribed on a sphere centred at *O*. The various formulae relating the parts of a spherical triangle can readily be manipulated by any student familiar with the nomenclature of plane trigonometry once he has grasped that the sides of a spherical triangle, being arcs of circles, must be expressed in angular, and not in linear, measure. Thus, in the triangle *ABC* of Fig. 369, the sides *a*, *b*, *c* are

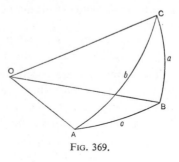

FIG. 369.

measured by the angles *BOC*, *COA*, *AOB* respectively, which they subtend at the centre of the sphere; where *A*, *B*, *C* are the poles of crystal faces, the sides *a*, *b*, *c* are the normal interfacial angles measured in optical goniometry. The angle *C* of the triangle is measured by the angle between the tangents at *C* to the arcs *AC* and *BC*; that is, it is the dihedral angle between the planes *AOC* and *BOC*. If the arcs *AC* and *BC* are portions of zone-circles, the angle *C* is the angle between two zone-axes. It is therefore the angle between two possible crystal edges and can be represented as a plane angle of a crystal face.

THE ESSENTIAL FORMULAE FOR CRYSTALLOGRAPHIC CALCULATIONS

If any three parts of a spherical triangle are of known magnitude, formulae can be derived to determine the values of the remaining three parts. (Remember that the sum of the three angles of a spherical triangle is not constant, and that three angles define a spherical triangle completely, since the radius of the sphere is immaterial.) We shall use the following formulae in our calculations:

To determine a side in terms of the remaining sides and the included angle.

$$\cos c = \cos a \cos b + \sin a \sin b \cos C. \quad\quad\dots\dots\dots\dots\dots(1)$$

To determine an angle in terms of the three sides.

$$\tan \frac{A}{2} = \sqrt{\frac{\sin(s-b)\sin(s-c)}{\sin s \sin(s-a)}} \quad \text{where} \quad s = \frac{a+b+c}{2}. \quad\dots\dots(2)$$

Formula (1) may also be rewritten

$$\cos C = \frac{\cos c - \cos a \cos b}{\sin a \sin b},$$

but this is not so convenient for logarithmic calculation.

To determine a side in terms of two angles and one other side, or to determine an angle in terms of two sides and an angle not included by them.

$$\frac{\sin a}{\sin A} = \frac{\sin b}{\sin B} = \frac{\sin c}{\sin C}. \quad\quad\dots\dots\dots\dots\dots\dots\dots\dots\dots(3)$$

The proof of these formulae is really outside the scope of crystallography, but we may illustrate the method of proof of formula (1), from which the others may be derived.

At C on the sphere (Fig. 370) draw CD, CE tangents to the arcs AC, BC at C. Since these tangents lie in the planes AOC, COB respectively, they will intersect the radii OA, OB at D and E. The plane angle ECD thus equals the angle C of the spherical triangle ABC.

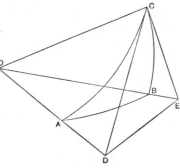

Fig. 370.

In $\triangle ECD$, $\quad DE^2 = CD^2 + CE^2 - 2CD \cdot CE \cos C,$

and in $\triangle EOD$, $\quad DE^2 = OD^2 + OE^2 - 2OD \cdot OE \cos c,$

and by subtraction:

$$0 = OD^2 - CD^2 + OE^2 - CE^2 + 2CD \cdot CE \cos C - 2OD \cdot OE \cos c.$$

Since the $\triangle$'s OCE, OCD are right-angled at C,

$$OD^2 - CD^2 = OC^2 = OE^2 - CE^2.$$

Therefore $\quad 0 = 2OC^2 + 2CD \cdot CE \cos C - 2OD \cdot OE \cos c;$

$$\cos c = \frac{OC}{OE} \frac{OC}{OD} + \frac{CE}{OE} \frac{CD}{OD} \cos C$$

$$= \cos a \cos b + \sin a \sin b \cos C.$$

THE SOLUTION OF RIGHT-ANGLED (NAPIERIAN) TRIANGLES

Though these general formulae are ultimately essential for calculations in the less symmetrical systems, we can often effect a great simplification by using for calculations in the more regular systems spherical triangles in which one part (whether an angle or a side) is a right-angle. The simplified formulae could be derived from formulae 1, 2, 3 above; thus, if $C = 90°$,

$$\cos c = \cos a \cos b, \quad \text{from (1)},$$

$$\sin a = \sin A \sin c, \quad \text{from (3), etc.}$$

It is tiresome, however, to need to remember six distinct, yet rather similar, formulae, and Napier long ago described a device by which the required formula can be written down at sight. Since these simplified formulae and the device for their derivation apply only to right-angled triangles, such triangles are often termed *Napierian triangles*. In Fig. 371 the five parts of the triangle (excluding the right-angle) are numbered in order, as they are encountered by moving around the triangle from the right-angle. On the right of the figure is drawn a five-compartment diagram, three compartments on the left of the vertical stroke, each filled in with the symbol $90° -$, and two on the right. This device must be sketched by the student *every time* a Napierian triangle is to be solved; the numbered parts of the triangle are then written in the appropriate compartments, starting from the horizontal line as we start from the right-angled part of the triangle, and proceeding from compartment to compartment in the same sense as we proceed around the triangle. Since the triangle is soluble, two of the parts 1, 2, 3, 4, 5 are already of known magnitude; any other part which we may require must be situated *either* in such a way that it and the two known parts

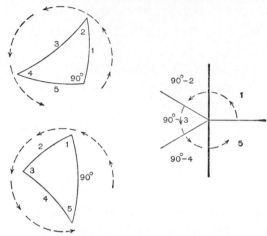

FIG. 371. Napier's device for the solution of right-angled spherical triangles.

are all three in adjacent compartments, *or* in such a way that two of the compartments in question are opposite the third.   All the required formulae are then summarised in the two statements:

The sine of a middle part = $\begin{cases}$ The product of the tangents of adjacent parts, $\\$ or $\\$ The product of the cosines of opposite parts. $\end{cases}$

Fig. 372 (*a*) shows an adjacent arrangement, with the middle part ringed, whilst in Fig. 372 (*b*) the arrangement is that of a middle part

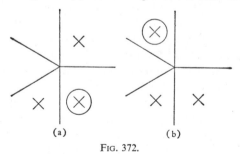

(a)   (b)

FIG. 372.

and two opposites.   (Note that it is not necessary that the unknown should be the middle part of the formula; an intermediate auxiliary solution is therefore never necessary, and we can always go straight to the required answer for any of the three unknown parts.)

THE CALCULATION OF AXIAL ANGLES AND AXIAL RATIOS

One of the earliest occasions on which the need for calculations of spherical triangles may arise follows directly after the measurement of a crystal. Having made a complete projection, determined the symmetry, selected the appropriate set of crystallographic axes and a parametral plane, we next require to calculate the constants for the substance—in the most general case the three axial angles α, β, γ and the axial ratios a/b and c/b. Any symmetry revealed by the projection enables us to average the actual measured values of angles thus seen to be theoretically equal, and these averaged values are used in calculation.

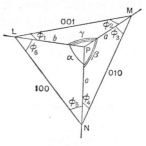

FIG. 373.

The general problem of a triclinic crystal may be studied first. In Fig. 373 there is portrayed the upper front right-hand corner of a crystal determined by the planes 100, 010 and 001, and across the corner is drawn the trace of the parametral plane *LMN*. Since the edges 010–001, etc., have been chosen as the directions of the crystallographic axes, the axial angles α, β, γ are the plane angles of the faces 100, 010, 001, so marked in the figure. The parametral plane cuts the three axial edges in lengths proportional to a, b and c. In the corresponding projection, Fig. 374, the primitive zone 100–010 is normal to the z-axis, the zone 010–001 is normal to the x-axis, and the zone 100–001 is normal to the y-axis. The axial angles α, β, γ are therefore given by the angles between these zones and are readily seen to be the supplements of the angles A, B, C of the triangle ABC.

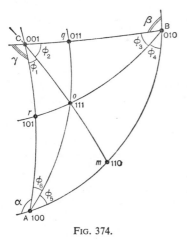

FIG. 374.

From the plane triangles of Fig. 373, we can derive the relationships

$$\frac{a}{b}=\frac{\sin \phi_1}{\sin \phi_2}, \quad \frac{c}{b}=\frac{\sin \phi_6}{\sin \phi_5}, \quad \frac{c}{a}=\frac{\sin \phi_3}{\sin \phi_4}.$$

The angle ϕ_1 is the acute angle between the edge 100–001 and the edge 111–001; it is therefore the acute angle between the corresponding zones on the stereogram, the angle ACm so marked in the projection. The other ϕ angles can be identified in the projection in a similar manner. We shall now proceed to apply these formulae to calculations in the various systems in turn.

In the *cubic system*, α, β, and γ each $=90°$, and the axial ratios are unity (the auxiliary ϕ angles each $=45°$).

In the *tetragonal system*, $\alpha = \beta = \gamma = 90°$. Also $\phi_1 = \phi_2 = 45°$, whence $a/b = 1$. The ratio $c/a\,(=c/b)$ is required. From Fig. 375,

$$\phi_6 = 001 \frown 011 = 90° - \phi_5;$$

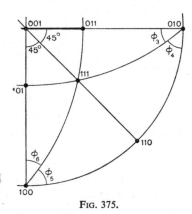

Fig. 375.

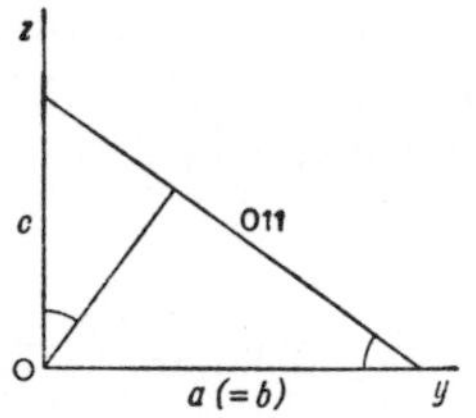

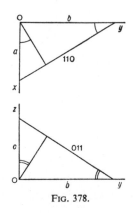

Fig. 376.

whence the axial ratio $=\tan 001 \frown 011$, a result easily derived also from simple geometrical considerations, for 011 is a plane cutting the orthogonal z and y axes in the ratio c/b (Fig. 376).

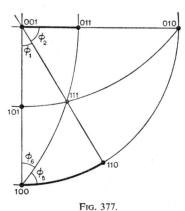

Fig. 377.

Fig. 378.

In the *orthorhombic system*, the axial angles are again 90°. By either of the above methods of reasoning (Figs. 377, 378),

$$a/b = \tan 100 \frown 110, \qquad c/b = \tan 001 \frown 011.$$

In the *monoclinic system*,

$$\alpha = \gamma = 90°, \quad \beta = 180° - (\phi_3 + \phi_4) = 180° - 100 \frown 001 \text{ (Fig. 379)}$$

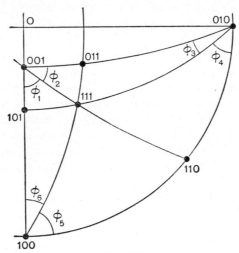

FIG. 379.

$$a/b = \frac{\sin \phi_1}{\sin \phi_2} = \tan \phi_1, \qquad c/b = \frac{\sin \phi_6}{\sin \phi_5} = \tan \phi_6;$$

also

$$c/a = \frac{\sin \phi_3}{\sin \phi_4} = \frac{\sin 001 \frown 101}{\sin 100 \frown 101}.$$

Alternative simple geometrical reasoning follows a slightly different course in deriving values for a/b and c/b, since the angles ϕ_1 and ϕ_6 are not directly related to interfacial angles derived from goniometric measurements (note particularly that, since 001 is not normal to faces in the primitive, ϕ_1 is *not* equal to the angle $100 \frown 110$). In Fig. 380 the tangent of the angle $100 \frown 110$ can be expressed in terms of the crystal constants, but the x axis is now no longer in the plane of the section drawn—it is inclined at an angle $\beta - 90°$ below the plane of the paper. The intercept made by the trace

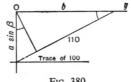

FIG. 380.

of the face 110 on the normal to 100 is thus no longer a, but $a \cos (\beta - 90°)$, i.e. $a \sin \beta$, whence

$$\tan 100 \frown 110 = \frac{a \sin \beta}{b}, \quad \text{.........................(1)}$$

and similarly $\tan 001 \frown 011 = \frac{c \sin \beta}{b}. \quad \text{.........................(2)}$

(By writing down a Napierian solution for the triangle $100 - 110 - 001$ the student should verify that the relationship (1) above is equivalent to the statement $a/b = \tan \phi_1$.)

Of the edges of a monoclinic crystal which are selected to determine the directions of the crystallographic axes, only those parallel to y are specially related to the symmetry elements. A prominent zonal direction is then selected as the z direction and set vertically; a form indexed $\{110\}$ is therefore very frequently present in practical problems on actual crystals, and a/b may be calculated by formula (1) above. It does not follow, however, that the form $\{011\}$ is equally likely to be present, and instead forms $\{h\,0\,l\}$ are often prominent. It is then easier to calculate c/a from consideration of the face 101 (or any $h\,0\,l$ or $\bar{h}\,0\,l$ face) and to derive c/b ultimately from the product $c/a \times a/b$. The geometrical derivation of c/a from the plane 101 is illustrated in Fig. 381. The x and z axes intersect at the obtuse angle β; from the triangle AOC formed by the axes and the trace of the face 101,

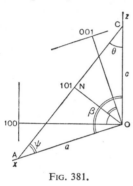

FIG. 381.

$$\frac{c}{a} = \frac{\sin \psi}{\sin \theta} = \frac{\sin 001 \frown 101}{\sin 100 \frown 101}. \quad \text{.........................(3)}$$

In the *triclinic system*, all three axial angles α, β, γ must be calculated directly from the triangle ABC (Fig. 374), since they are not simply related to the interfacial angles measured on the goniometer. If a face such as 110 is also present, the corresponding axial ratio can be expressed in terms of measured interfacial angles, avoiding the necessity for actual calculation of the auxiliary angles ϕ_1 and ϕ_2. In Fig. 374,

from the triangle ACm, $\dfrac{\sin \phi_1}{\sin Am} = \dfrac{\sin \overset{\wedge}{CmA}}{\sin AC}$, and from the triangle BCm, $\dfrac{\sin \phi_2}{\sin Bm} = \dfrac{\sin (180° - \overset{\wedge}{CmA})}{\sin BC}$;

whence

$$\frac{a}{b} = \frac{\sin \phi_1}{\sin \phi_2} = \frac{\sin Am \cdot \sin BC}{\sin Bm \cdot \sin AC}$$

$$= \frac{\sin 100 \frown 110 \cdot \sin 010 \frown 001}{\sin 010 \frown 110 \cdot \sin 100 \frown 001} \, .$$

In the *hexagonal system*, we proceed most easily by a simple geo-

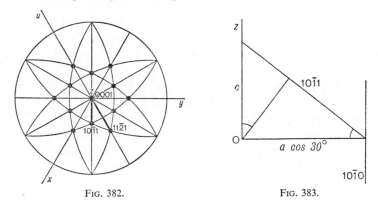

Fig. 382.   Fig. 383.

metrical derivation. From the projection, Fig. 382, and the diagram, Fig. 383, it is readily apparent that $\tan 0001 \frown 10\bar{1}1 = \dfrac{c}{a \cos 30°}$. Alternatively, one may calculate from the position of the pole $11\bar{2}1$, for $\tan 0001 \frown 11\bar{2}1 = \dfrac{c}{a/2}$, whence $c/a = \frac{1}{2} \tan 0001 \frown 11\bar{2}1$ (Fig. 384).

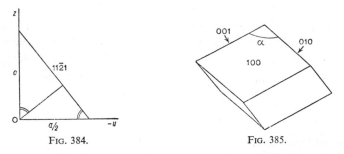

Fig. 384.   Fig. 385.

In the *trigonal system*, using Miller-Bravais notation, the calculation of the axial ratio, of course, proceeds precisely as in the hexagonal system. In Miller's three-index notation the axial units are all equal,

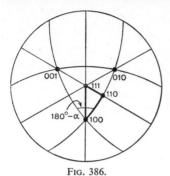

FIG. 386.

and we require to calculate the angle α, between the axes, characteristic of the substance. Since the zones 100–001 and 100–010 are normal to the rhombohedron edges including the angle α (Fig. 385), we require to calculate the angle between these zones on the stereogram (Fig. 386). If the interfacial angle of the rhombohedron {100} has been measured goniometrically, the triangle marked on the stereogram is soluble by a Napierian solution to give one-half of the supplement of the required angle α.

SOME EXAMPLES OF THE CALCULATION OF AXIAL RATIOS

Tetragonal system.

In mercurous chloride (p. 62), the average value of the angle $100 \frown 111$ was determined as $49° 5'$. We require to solve a triangle for the angle $001 \frown 011$ (Fig. 387); since the crystal is tetragonal, the

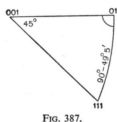

FIG. 387.

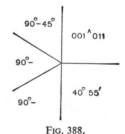

FIG. 388.

angle of the triangle at $001 = 45°$, and the side $111 \frown 011 = 90° - 49° 5'$. The angle of the triangle at $011 = 90°$, and we can derive a Napierian solution. Sketching Napier's device (Fig. 388), we can select the

equation $\sin 001 \frown 011 = \tan 45° \tan 40° 55'$,

whence $001 \frown 011 = 60° 5'$

and $c/a = \tan 60° 5' = \underline{1{\cdot}738}.$

Orthorhombic system.

In potassium sulphate, the following average values were obtained as the result of measurement of several crystals:

$$100 \frown 111 = 43° 52', \quad 001 \frown 111 = 56° 11'.$$

The values are inserted in a quadrant of the stereogram (Fig. 389), and we draw the Napierian device for the triangle outlined (Fig. 390).

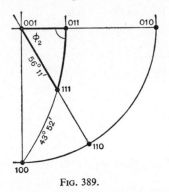

FIG. 389.

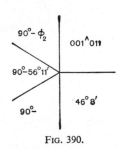

FIG. 390.

Writing down the appropriate solutions:

$\sin 46° 8' = \sin \phi_2 \sin 56° 11'$, $\quad \cos 56° 11' = \cos 001 \frown 011 \cos 46° 8'$,

whence $\phi_2 = 60° 12'$, $\qquad\qquad$ whence $001 \frown 011 = 36° 34'$,

$\dfrac{a}{b} = \cot \phi_2 = \underline{0·573}.$ $\qquad\qquad$ $\dfrac{c}{b} = \tan 36° 34' = \underline{0·742}.$

Monoclinic system.

(1) The following average values were obtained by measurement of crystals of lead chromate:

$$1\bar{1}0 \frown 110 = 86° 19',$$
$$110 \frown 001 = 80° 57',$$
$$010 \frown 011 = 48° 13'.$$

The triangle outlined in Fig. 391 is first solved for the value of the

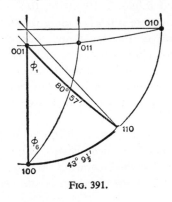

FIG. 391.

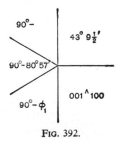

FIG. 392.

side $100 \frown 001$, giving the supplement of the required axial angle β. From Napier's device (Fig. 392) we write down

$$\cos 80° 57' = \cos 43° 9\tfrac{1}{2}' \cos 001 \frown 100,$$

whence $\qquad\qquad 001 \frown 100 = 77° 33'$

and $\qquad\qquad\qquad \beta = \underline{102° 27'}.$

The axial ratio $\qquad \dfrac{a}{b} = \dfrac{\tan 100 \frown 110}{\sin \beta}$

$$= \frac{\tan 43° 9\tfrac{1}{2}'}{\sin 102° 27'}$$

$$= \underline{0.960}.$$

Alternatively, we might solve for the angle ϕ_1 from the Napierian device,

$$\sin 43° 9\tfrac{1}{2}' = \sin 80° 57' \sin \phi_1,$$

whence $\qquad\qquad\qquad \phi_1 = 43° 50'$

and $\qquad\qquad\qquad \dfrac{a}{b} = \tan 43° 50'$

$$= 0.960.$$

The axial ratio $\qquad \dfrac{c}{b} = \dfrac{\tan 001 \frown 011}{\sin \beta}$

$$= \frac{\tan 41° 47'}{\sin 102° 27'}$$

$$= \underline{0.915}.$$

We have thus determined the constants for lead chromate:

$$\beta = 102° 27'; \quad a:b:c = 0.960:1:0.915.$$

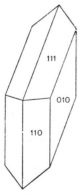

FIG. 393. Measured crystal of gypsum.

(2) The forms present on the crystal of lead chromate considered above—{110}, {010}, {011}, {001} —enabled us to measure angles leading directly to a simple calculation of the constants. Frequently the habit of a monoclinic crystal is such that the subsequent calculations are less straightforward, and each case must be considered on its own merits. As a typical example we shall calculate the constants for gypsum.

The commonest habit is illustrated in Fig. 393, on which are marked also the indices usually assigned to these forms; notice that the downward sloping edges have not been chosen as the direction of the x-axis, the sloping planes being indexed {111}. The interfacial angles obtained by measurement are:

$$010 \wedge 110 = 55° 45',$$
$$010 \wedge 111 = 71° 54',$$
$$110 \wedge 111 = 49° 9'.$$

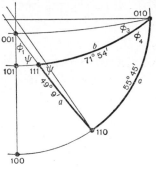

These three angles give the values of the three sides of a non-Napierian (or oblique) spherical triangle (Fig. 394), and we must clearly begin by solving this triangle for one or more of its angles. We shall solve for ϕ_4 ($= 100 \wedge 101$), and also for the

Fig. 394.

auxiliary angle ψ; the most convenient formula is number (2), p. 182. The solution is set out in detail below.

		log sin
$a = 49° 9'$,	$s - a = 39° 15'$,	$\bar{1}\cdot 8012$,
$b = 71° 54'$,	$s - b = 16° 30'$,	$\bar{1}\cdot 4533$,
$c = 55° 45'$,	$s - c = 32° 39'$,	$\bar{1}\cdot 7320$,
2 ⌊176° 48',		
$s = 88° 24'$.	$s = 88° 24'$.	$\bar{1}\cdot 9998$.

$$\log \tan \frac{A}{2} = \tfrac{1}{2}(\bar{1}\cdot 1853 - \bar{1}\cdot 8010) \qquad \log \tan \frac{C}{2} = \tfrac{1}{2}(\bar{1}\cdot 2545 - \bar{1}\cdot 7318)$$

$$= \bar{1}\cdot 6921 \qquad\qquad\qquad = \bar{1}\cdot 7613$$

$$\frac{A}{2} = 26° 12', \qquad\qquad\qquad \frac{C}{2} = 30° 0',$$

$$\phi_4 = A = 52° 24'. \qquad\qquad\qquad \psi = 60° 0'.$$

We can now use the value of ψ in a Napierian solution of the triangle

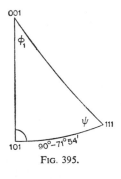

Fig. 395.

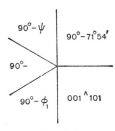

Fig. 396.

001 – 101 – 111 (Fig. 395).   From the Napierian diagram (Fig. 396) we write down:

$$\cos 71° \ 54' = \cot 60° \ 0' \tan 001 \frown 101, \qquad \cos \phi_1 = \sin 60° \ 0' \sin 71° \ 54',$$

whence $001 \frown 101 = 28° \ 17'$, $\qquad\qquad$ whence $\phi_1 = 34° \ 36'$,

$$180° - \beta = 100 \frown 101 + 101 \frown 001$$
$$= 52° \ 24' + 28° \ 17',$$
$$\beta = \underline{99° \ 19'}.$$

$$\frac{a}{b} = \tan \phi_1$$
$$= \underline{0 \cdot 690}.$$

The axial ratio $\dfrac{c}{a} = \dfrac{\sin 001 \frown 101}{\sin 100 \frown 101}$

$$= \frac{\sin 28° \ 17'}{\sin 52° \ 24'},$$

whence $\qquad \dfrac{c}{b} = \dfrac{c}{a} \times \dfrac{a}{b} = \dfrac{0 \cdot 690 \times \sin 28° \ 17'}{\sin 52° \ 24'}$

$$= \underline{0 \cdot 412}.$$

The crystallographic constants of gypsum in this setting are thus:

$$\beta = 99° \ 19'; \quad a : b : c = 0 \cdot 690 : 1 : 0 \cdot 412.$$

Triclinic system.

In basic mercurous nitrate, $3HgNO_3 \cdot 2HgOH$, the following measurements were made:

$$100 \frown 110 = 42° \ 58' \qquad 010 \frown 011 = 37° \ 29'$$
$$\qquad\qquad\qquad\qquad\qquad\qquad\qquad\qquad\qquad 100 \frown 001 = 69° \ 1'$$
$$110 \frown 010 = 33° \ 33' \qquad 011 \frown 001 = 44° \ 3'$$

To determine the axial angles, we must solve for all three angles of the oblique triangle ABC (Fig. 397).

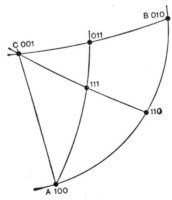

FIG. 397.

$$\begin{array}{llr}
& & \text{log sin} \\
a = 81° \ 32', & s-a = 32° \ 0', & \bar{1}\cdot7242. \\
b = 69° \ 1', & s-b = 44° \ 31', & \bar{1}\cdot8458. \\
c = 76° \ 31', & s-c = 37° \ 1', & \bar{1}\cdot7796. \\
2 \ \underline{|\ 227° \ 4'} & & \\
s = 113° \ 32'. & s = 113° \ 32'. & \bar{1}\cdot9623.
\end{array}$$

$$\log \tan \frac{A}{2} = \tfrac{1}{2}(\bar{1}\cdot6254 - \bar{1}\cdot6865) \qquad \log \tan \frac{B}{2} = \tfrac{1}{2}(\bar{1}\cdot5038 - \bar{1}\cdot8081)$$

$$= \bar{1}\cdot9694. \qquad\qquad\qquad\qquad = \bar{1}\cdot8478.$$

$$\frac{A}{2} = 42° \ 59'. \qquad\qquad\qquad\qquad \frac{B}{2} = 35° \ 10'.$$

$$\alpha = 180° - 85° \ 58' \qquad\qquad\qquad \beta = 180° - 70° \ 20'$$

$$= \ 94° \ 2'. \qquad\qquad\qquad\qquad = 109° \ 40'.$$

$$\log \tan \frac{C}{2} = \tfrac{1}{2}(\bar{1}\cdot5700 - \bar{1}\cdot7419)$$

$$= \bar{1}\cdot9140.$$

$$\frac{C}{2} = 39° \ 22'.$$

$$\gamma = 180° - 78° \ 44'$$

$$= 101° \ 16'.$$

The axial ratio $\dfrac{a}{b} = \dfrac{\sin 42° \ 58' \ \sin 81° \ 32'}{\sin 33° \ 33' \ \sin 69° \ 1'}$ (p. 189)

$$= 1\cdot306,$$

and the axial ratio $\dfrac{c}{b} = \dfrac{\sin 44° \ \ 3' \ \sin 76° \ 31'}{\sin 37° \ 29' \ \sin 69° \ \ 1'}$

$$= 1\cdot190.$$

We have thus determined the constants

$$\alpha = 94° \ 2'; \quad \beta = 109° \ 40'; \quad \gamma = 101° \ 16'; \quad a : b : c = 1\cdot306 : 1 : 1\cdot190.$$

The trigonal system.

A single example of the calculation of an axial angle in Miller's notation will suffice. In calcite, the cleavage angle $100 \frown 010 = 74° \ 55'$ (Fig. 398); the Napierian triangle of Fig. 399 is soluble for one-half of the supplement of α. From the Napierian diagram (Fig. 400):

$$\cos 60° = \sin \frac{\theta}{2} \cos 37° \ 27\tfrac{1}{2}',$$

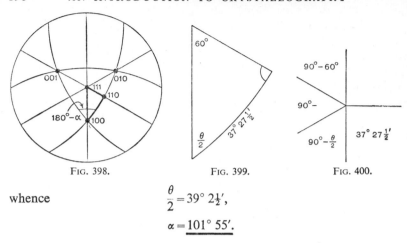

| FIG. 398. | FIG. 399. | FIG. 400. |

whence

$$\frac{\theta}{2} = 39° \ 2\tfrac{1}{2}',$$

$$\alpha = 101° \ 55'.$$

THE CONSTRUCTION OF AN AXIAL CROSS FOR CRYSTAL-DRAWING

If an investigation is to be illustrated by drawing a typical crystal on a set of axes, the clinographic projection of the three equal orthogonal axes of the cubic system (p. 46) must be modified in each of the other systems to accord with the values of axial angles and axial ratios calculated as we have shown above.

In the *tetragonal system* the x and y axes remain unaltered, but the z axis must be modified to agree with the calculated ratio c/a. The actual length of the z axis in the drawing of cubic axes (which represents, of course, the same length as the x and y axes in space) is measured on a convenient scale, multiplied by the ratio c/a for the particular substance, and the new position of $\pm z$ plotted on the axial cross.

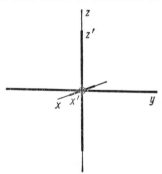

FIG. 401. Cubic axial cross modified to accord with the axial ratios of potassium sulphate.

In the *orthorhombic system* the z axis is modified as in the tetragonal system; the apparent length of the x axis of the cubic cross is likewise measured, and this also is modified in accordance with the calculated ratio a/b. Fig. 401 represents the correct axial cross for potassium sulphate, for which we have determined

$$a : b : c = 0.573 : 1 : 0.742.$$

In the *monoclinic system* the unit-length x axis is first rotated in the xz axial plane to make the appropriate angle β with the z axis. A plan of the necessary construction on the xz axial plane is shown in Fig. 402; a point M is found on Ox such that $OM = Ox' \sin \beta = Ox \sin \beta$, and a point N on the negative z axis such that $ON = Ox' \cos \beta = Oz \cos \beta$. The required new position Ox' of the x axis is then the diagonal of the rectangle $MONx'$. Adapting this construction to the clinographic representation of the cubic axial cross, the point M (Fig. 403) must be located by measuring the apparent length of Ox as represented on the paper and multiplying this by $\sin \beta$; N is similarly located on the negative z axis, measuring the apparent length of the z unit distance and multiplying this by $\cos \beta$. Mx' and Nx' are then drawn parallel to the original axes, to meet at x'. Then Ox' represents the new axis, still the same length as Oy and Oz in space but sloping at the correct angle β. Finally, the lengths Ox' and Oz on the paper are modified to accord with the calculated ratios a/b and c/b.

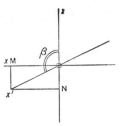

Fig. 402. Plan on the xz plane of the rotation of the x-axis to accord with a particular value of β.

In the *triclinic system* we again proceed by first constructing a set of axes of unit length correctly inclined at the calculated angles α, β, γ.

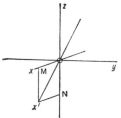

Fig. 403. Construction for rotation of the x-axis on the axial cross.

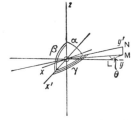

Fig. 404. Construction of a triclinic axial cross.

The x axis is moved downwards in the original xz plane to the correct β value precisely as in the monoclinic system. The y axis must be moved back and up (or down) to make both γ and α agree with the calculated values. This is accomplished by means of co-ordinates OL, LM, MN (Fig. 404), measured parallel to the original y, x and z directions respectively.

$MN = ON \cos \alpha = \cos \alpha \times$ original length of Oz on the paper,

$ML = OM \cos \theta = ON \sin \alpha \cos \theta$

$\qquad = \sin \alpha \cos \theta \times$ original length of Ox on the paper,

$OL = OM \sin \theta = ON \sin \alpha \sin \theta$

$\qquad = \sin \alpha \sin \theta \times$ original length of Oy on the paper,

where $\theta =$ angle OML.

Moreover, the crystal face 010 is parallel to the axial plane $x'z$ and therefore is parallel to the plane NML;

whilst the crystal face 100 is parallel to the axial plane yz and therefore is parallel to the plane NMO,

whence θ is the *interfacial* angle 100⌒010, and we can thus determine the values of the co-ordinates OL, LM, MN. Having thus constructed a set of unit axes at the required angles α, β, γ, the x' and z unit lengths are finally modified to agree with the calculated axial ratios a/b and c/b.

In the *hexagonal system* the y axis is retained in its original position, and new x and u axes are constructed, still in the plane normal to z but making angles of 120° with y. Seen in a plan on the xy plane this involves the construction illustrated in Fig. 405. The original x

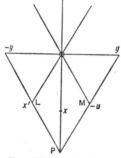

Fig. 405. Plan on the xy plane of the construction of a Miller-Bravais axial cross.

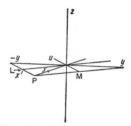

Fig. 406. Construction of a Miller-Bravais axial cross

axis is produced to a point P, so that $OP = Oy \tan 60° = 1 \cdot 73 Oy = 1 \cdot 73 \times$ original length of the x axis. P is then joined to $\pm y$, and L, M are the middle points of Py, $P\bar{y}$. Joining L and M to the origin gives the new $+x$ and $-u$ axes. On the clinographic axes (Fig. 406) the apparent length of Ox as measured on the paper is multiplied by $1 \cdot 73$ to give the point P, and the points L and M are found as before. In drawing a particular substance, the z axis is modified in the usual way to agree with the calculated axial ratio c/a.

In the *trigonal system* earlier crystallographers developed special

methods for drawing crystals. These have now been completely superseded, and if a drawing is to be made on an axial cross, Miller indices
are first converted to the corresponding Miller-Bravais symbols, the
appropriate axial ratio is calculated and the drawing is made on the
corresponding hexagonal axes. Drawings are often made, however,
directly from the projection by methods to be described later (*v.*
Chapter IX).

THE EQUATIONS OF A NORMAL

The crystallographer usually defines the slope of a crystal face in
terms of its Miller indices $h\,k\,l$, but the same result might be achieved
by stating the angles which the normal to the
face makes with the crystallographic axes (the
student familiar with analytical methods will
recognise here the use of *direction-cosines*). In
Fig. 407 the plane $h\,k\,l$ cuts the axes at the points

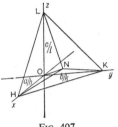

$H,\ K,\ L$; thus if $OH=\dfrac{a}{h}$, $OK=\dfrac{b}{k}$ and $OL=\dfrac{c}{l}$.

Draw ON normal to the plane from the origin.
Then

Fig. 407.

$$\cos N\hat{O}H=\frac{ON}{a/h},\quad \cos N\hat{O}K=\frac{ON}{b/k},\quad \cos N\hat{O}L=\frac{ON}{c/l}.$$

Writing $N\hat{O}H$ as $N\frown x$, etc.,

$$\frac{a}{h}\cos N\frown x=\frac{b}{k}\cos N\frown y=\frac{c}{l}\cos N\frown z,$$

which expressions are known as the equations of the normal. They are
true in any system, whether the crystallographic
axes are orthogonal or not, but are mostly used
in calculations in the cubic, tetragonal and orthorhombic systems. In these systems, each of the
axes x, y, z is normal to a possible crystal face,
and the values of the angles $N\frown x$, $N\frown y$, $N\frown z$
are directly related to angles measurable on the
goniometer.

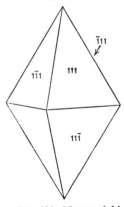

Thus, in orthorhombic sulphur the following
angles were measured on the bipyramid (Fig. 408)
chosen as the parametral form:

$$111\frown \bar{1}11 = 94° \ 52',$$
$$111\frown 1\bar{1}1 = 73° \ 34',$$
$$111\frown 11\bar{1} = 36° \ 40\tfrac{1}{2}'.$$

Fig. 408. Measured bi-
pyramidal crystal of sul-
phur.

From the stereogram (Fig. 409) it is readily seen that

$$N \frown x = 90° - \tfrac{1}{2}(94° \ 52') \ = 42° \ 34',$$
$$N \frown y = 90° - \tfrac{1}{2}(73° \ 34') \ = 53° \ 13',$$
$$N \frown z = 90° - \tfrac{1}{2}(36° \ 40\tfrac{1}{2}') = 71° \ 39\tfrac{3}{4}'.$$

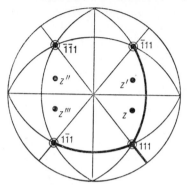

FIG. 409. Stereogram of a sulphur crystal.

Writing down the equations of the normal:

$$\frac{a}{1} \cos 42° \ 34' = \frac{b}{1} \cos 53° \ 13' = \frac{c}{1} \cos 71° \ 39\tfrac{3}{4}',$$

$$\frac{a}{0·813} = \frac{b}{1} = \frac{c}{1·903},$$

whence the axial ratios $a : b : c = 0·813 : 1 : 1·903$.

On a more highly-modified crystal, measurements were made on a further bipyramidal form, z (Fig. 409).

$$zz' \ = \ 34° \ 17',$$
$$zz'' = 101° \ 58',$$
$$zz''' = \ 91° \ 50'.$$

What is the index of the form z?

Let it be $h \, k \, l$.   Then

$$\frac{0·813}{h} \cos 72° \ 51\tfrac{1}{2}' = \frac{1}{k} \cos 44° \ 5' = \frac{1·903}{l} = \cos 50° \ 59'$$

$$\frac{0·333}{h} = \frac{1}{k} = \frac{1·668}{l},$$

$$\frac{1}{h} = \frac{3}{k} = \frac{5}{l},$$

and the form z is {135}.

It may be useful to insert here a note on the habit of referring to a given crystal form by a letter, as in the form $z\{135\}$ above. Such lettering is a great convenience in crystal descriptions and tabulations, for it avoids the necessity for constantly repeating the full index. Haüy lettered the faces of his 'primitive forms' P, M, and T from PriMiTive, and a systematic use of lettering is fairly widely observed in French literature. Crystallographers in other countries have unfortunately never adopted either the French or any other scheme uniformly and the ultimate authority for the use of a particular letter for a particular form is the author of the original description, or a recognised work of reference such as Groth's *Chemische Krystallographie*. The following are fairly generally used: $a\{100\}$, $b\{010\}$, $c\{001\}$ (or capital letters A, B, C to avoid confusion with axial units); $m\{110\}$, $o\{111\}$; $M\{1\bar{1}0\}$ where this is a different form from $m\{110\}$. In the hexagonal system $c\{0001\}$, $m\{10\bar{1}0\}$, $a\{11\bar{2}0\}$, $s\{11\bar{2}1\}$. In the trigonal system $c\{111\}$, $r\{100\}$, $a\{1\bar{1}0\}$, $e\{110\}$. Individual faces of a form may be distinguished by apostrophes, as $a(100)$, $a'(\bar{1}00)$.

In the monoclinic and triclinic systems, some or all of the angles $N\frown x$, $N\frown y$, $N\frown z$ are no longer measured by the crystallographic interfacial angles $h\,k\,l\frown100$, $h\,k\,l\frown001$, and the equations to the normal are less directly useful. They may, however, be expressed in terms of goniometrical angles. In the monoclinic system, for example

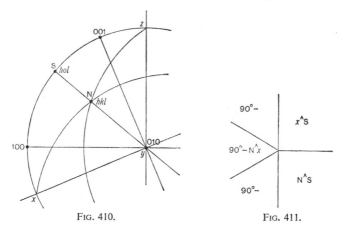

FIG. 410.   FIG. 411.

(Fig. 410), we can write down a Napierian solution for the triangles $x\,NS$, $z\,NS$. From the appropriate diagram (Fig. 411),

$$\cos N\frown x = \cos x\frown S \cos N\frown S;$$

and, similarly, $\cos N\frown z = \cos z\frown S \cos N\frown S$.

From the equations to the normal,

$$\frac{a}{h}\cos N\frown x = \frac{b}{k}\cos N\frown y = \frac{c}{l}\cos N\frown z,$$

whence

$$\frac{a}{h}\cos x\frown S \cos N\frown S = \frac{b}{k}\cos N\frown y = \frac{c}{l}\cos z\frown S \cos N\frown S,$$

but

$$x\frown S = 90° - h\,0\,l\frown 001,$$
$$N\frown S = 90° - N\frown y,$$
$$z\frown S = 90° - h\,0\,l\frown 100.$$

Therefore

$$\frac{a}{h}\sin h\,0\,l\frown 001 = \frac{b}{k}\cot N\frown y = \frac{c}{l}\sin h\,0\,l\frown 100,$$

in which form the equations may sometimes be useful in this system. (We have derived one of these equations already during our calculations of axial ratios.)

THE ANGLE BETWEEN TWO FACE-NORMALS

Having turned our attention to the possibility of determining a face-normal in terms of the angles $N\frown x$, $N\frown y$ and $N\frown z$, we may enquire whether this leads to a simple expression for the value of the angle between two such face-normals. The appropriate formulae are readily available in analytical solid geometry, but they are extremely cumbrous in the less symmetrical systems. In the triclinic system the expression for the value of the cosine of the angle θ between the normals to two faces $h_1 k_1 l_1$ and $h_2 k_2 l_2$ involves the axial angles α, β, γ, the axial units a, b, c, and the two sets of indices. Even with the simplification introduced by the orthogonal axes of the orthorhombic system the required expression reads:

$$\cos\theta = \frac{h_1 h_2 \dfrac{bc}{a} + k_1 k_2 \dfrac{ca}{b} + l_1 l_2 \dfrac{ab}{c}}{\sqrt{h_1{}^2 \dfrac{bc}{a} + k_1{}^2 \dfrac{ca}{b} + l_1{}^2 \dfrac{ab}{c}}\sqrt{h_2{}^2 \dfrac{bc}{a} + k_2{}^2 \dfrac{ca}{b} + l_2{}^2 \dfrac{ab}{c}}},$$

and it is clear that the crystallographer's method of calculating directly by the solution of the appropriate spherical triangle is much more manageable than this approach.

In the *cubic system*, however, the further simplification introduced by equal axial units results in the formula

$$\cos\theta = \frac{h_1 h_2 + k_1 k_2 + l_1 l_2}{\sqrt{h_1{}^2 + k_1{}^2 + l_1{}^2}\sqrt{h_2{}^2 + k_2{}^2 + l_2{}^2}},$$

and in this form it is often useful. Thus, to calculate the angle between two triad axes in the cubic system, since these are normal to octahedral faces, e.g. 111, 1$\bar{1}$1,

$$\cos\theta = \frac{1-1+1}{\sqrt{3}\,\sqrt{3}}$$

$$=0{\cdot}3333,$$

whence $\theta = 70° \ 32',$

a solution which is even more rapid than the solution of the appropriate Napierian triangle, but it must be repeated that this simplified formula is *applicable only in the cubic system.*

THE ZONE SYMBOL

We have already defined a zone as a set of faces with mutually parallel intersections, this common direction being termed the zone-axis. To specify the direction of a zone-axis, we may regard it as a line drawn through the origin of the crystallographic axes and quote the co-ordinates of a point on the line. If the co-ordinates are U, V, W, the zone-axis is the diagonal of a parallelepiped of sides aU, bV, cW (Fig. 412). These co-ordinates UVW constitute the *symbol* of the particular zone in question; note especially that they are used as true co-ordinates, and are therefore multipliers of the axial units a, b, c, and not divisors like face-indices. They are distinguished from face-indices by enclosure in square brackets, or crotchets, as $[UVW]$;

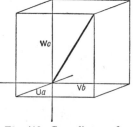

FIG. 412. Co-ordinates of a point on a zone axis.

to assist in keeping this important distinction in mind, it is best to refer to them always as *zone symbols* rather than as *zone indices*, a term used by some crystallographers. If the indices of two faces defining the zone-axis are $h_1k_1l_1$ and $h_2k_2l_2$, then it can be shown that

$$U = k_1l_2 - l_1k_2,$$

$$V = l_1 h_2 - h_1l_2,$$

$$W = h_1k_2 - k_1h_2.$$

A geometrical demonstration of these relationships is clumsy, and we shall not give one here. They are readily derived analytically; the equations of the planes through the origin, parallel to the crystal faces $h_1 k_1 l_1$ and $h_2 k_2 l_2$ respectively are

$$h_1 \frac{x}{a} + k_1 \frac{y}{b} + l_1 \frac{z}{c} = 0,$$

$$h_2 \frac{x}{a} + k_2 \frac{y}{b} + l_2 \frac{z}{c} = 0,$$

and the equations of the line at their intersection are therefore

$$\frac{x}{a(k_1 l_2 - l_1 k_2)} = \frac{y}{b(l_1 h_2 - h_1 l_2)} = \frac{z}{c(h_1 k_2 - k_1 h_2)}.$$

The numerical values of U, V, W for a given pair of faces are written down in practice by a convenient device known as *cross-multiplication* (the crystallographer's own adaptation of the determinant notation of the mathematician). Each index is written down twice, the second below the first. The first and last figures are crossed off, and the figures joined by the arms of each cross are multiplied, the product of a pair joined by a stroke downward to the left being subtracted from the product of the associated pair joined by a stroke downward to the right:

$$
\begin{array}{c|cccc|c}
h_1 & k_1 & l_1 & h_1 & k_1 & l_1 \\
 & & \times & \times & \times & \\
h_2 & k_2 & l_2 & h_2 & k_2 & l_2
\end{array}
$$

$$[k_1 l_2 - l_1 k_2, \quad l_1 h_2 - h_1 l_2, \quad h_1 k_2 - k_1 h_2]$$

If the numerical computation of a symbol $[UVW]$ has not been actually carried out, we can still symbolise a zone by enclosing in the characteristic square brackets the indices of a pair of faces in the zone
$-[(h_1 k_1 l_1), (h_2 k_2 l_2)]$.

It seems advisable to stress again here that since the figures making up a zone symbol are used as co-ordinates, whilst those composing the index of a face express sub-multiples of the axial units, there is no simple relationship between the two symbols $[pqr]$ and (pqr). The former represents the direction of an edge and the latter a crystal plane, and except through accidental relationships arising from the regular geometry of particular systems, the direction $[pqr]$ is *not* that of the normal to the face (pqr). Thus, if we calculate the zone symbol $[(100), (010)]$:

$$
\begin{array}{c|cccc|c}
1 & 0 & 0 & 1 & 0 & 0 \\
 & & \times & \times & \times & \\
0 & 1 & 0 & 0 & 1 & 0
\end{array}
$$

$$[001]$$

we obtain [001] which, from consideration of the faces which define it, must be the *symbol of the z crystallographic axis.* In the monoclinic and triclinic systems this direction [001] is clearly not normal to (001) Still less, then, will it be true in general that a symbol [$p\,q\,r$] represents a direction normal to ($p\,q\,r$).

THE WEISS ZONE LAW.

If any other face $h\,k\,l$ lies in the zone [UVW] defined by the faces $h_1 k_1 l_1$, $h_2 k_2 l_2$, then

$$Uh + Vk + Wl = 0 \quad \dots\dots\dots\dots\dots\dots\dots(1)$$

(for the equation to the plane through the origin parallel to ($h\,k\,l$) is

$$h\frac{x}{a} + k\frac{y}{b} + l\frac{z}{c} = 0, \quad \dots\dots\dots\dots\dots\dots(2)$$

and if the three planes are tautozonal this must contain the line

$$\frac{x}{aU} = \frac{y}{bV} = \frac{z}{cW}, \quad \dots\dots\dots\dots\dots\dots\dots(3)$$

and, by substituting in (2) the values of $\frac{x}{a}, \frac{y}{b}, \frac{z}{c}$ derived from (3), we arrive at equation (1)).

This important relationship, the zone law or zonal equation, was first enunciated by C. S. Weiss * (though he did not express it in the Millerian notation now used) and may conveniently be referred to as the *Weiss Zone Law.* It is the foundation of the ' adding rule ' by which we have found indices of other faces lying in a zone between two given faces (for it will be readily seen that any index of the type $mh_1 + nh_2$, $mk_1 + nk_2$, $ml_1 + nl_2$ must satisfy the Weiss equation).

In addition to its use for verifying the tautozonal relationship of given faces, the zone law also provides the means by which we can locate on a projection a zone which is referred to by its symbol. Suppose, for example, we wish to locate the zone [$1\bar{1}\bar{2}$]. By the zone law, for all faces $h\,k\,l$ in this zone,

$$h - k - 2l = 0.$$

Hence,　if $l = 0$ then $h = k$, and 110 is a face in the zone；
　　and if $k = 0$ then $h = 2l$, and 201 is a face in the zone.

* Christian Samuel Weiss was born in Leipzig in 1780. After studying mineralogy under Werner at Freiberg he worked also in Vienna and Paris. He first drew attention to the zone law in a translation into German of Haüy's treatise on mineralogy. His name is usually associated with a crystallographic notation involving the actual parameters of a face instead of sub-multiples of these, a notation which has scarcely survived to the present time, but he did in fact anticipate Whewell and Miller in using also our modern index notation. In 1810 he was appointed Professor of Mineralogy and Director of the Mineralogical Museum in Berlin, where he died in 1856.

The trace of the required zone-circle can thus be drawn as the great circle passing through the poles (110) and (201).

INDEX OF A FACE AT THE INTERSECTION OF TWO ZONES

If a face $(h\,k\,l)$ lies at the intersection of two zones $[UVW]$ and $[U'V'W']$, then h, k, l must satisfy the equations

$$Uh + Vk + Wl = 0,$$
$$U'h + V'k + W'l = 0,$$

and we require to solve these equations for the ratios $h : k : l$. This is easily accomplished by the same device of cross-multiplication:

$$
\begin{array}{c|cccc|c}
U & V & W & U & V & W \\
 & \times & \times & \times & \\
U' & V' & W' & U' & V' & W'
\end{array}
$$
$$(VW' - WV', \quad WU' - UW', \quad UV' - VU')$$

the smooth brackets re-appearing to remind us that it is a *face-index* which we have calculated.

Suppose, for example, that we require to find the index of the face in which the zones [321, 021] and [331, 001] intersect. Calculating the symbol of the first zone:

$$
\begin{array}{c|cccc|c}
3 & 2 & 1 & 3 & 2 & 1 \\
 & \times & \times & \times & \\
0 & 2 & 1 & 0 & 2 & 1
\end{array}
$$
$$[0\bar{3}6]$$

and for the second zone:

$$
\begin{array}{c|cccc|c}
3 & 3 & 1 & 3 & 3 & 1 \\
 & \times & \times & \times & \\
0 & 0 & 1 & 0 & 0 & 1
\end{array}
$$
$$[3\bar{3}0]$$

Cross-multiplying these two symbols:

$$
\begin{array}{c|cccc|c}
0 & \bar{3} & 6 & 0 & \bar{3} & 6 \\
 & \times & \times & \times & \\
3 & \bar{3} & 0 & 3 & \bar{3} & 0
\end{array}
$$
$$(18 . 18 . 9)$$

and the required face is (221), the zones of course intersecting also in the parallel face $(\bar{2}\bar{2}\bar{1})$.

In carrying out zone-axis calculations in Miller-Bravais notation, one of the non-independent first three indices is omitted throughout (conveniently that for the u axis) and the resultant co-ordinates are of course referred to the three remaining axes. To indicate this omission, the symbol may be written $[UV \dagger W]$; we shall not attempt here to derive a meaning for a possible figure to insert in place of the $\dagger$, since the three-index symbol is complete in itself and can be used both in geometrical construction and in calculations such as those involving the zone law. The figure $-(U + V)$ must *not* be inserted in such a zone symbol. When, however, the calculation leads ultimately to a face index $(h\,k * l)$ the third figure can be re-inserted, since we know that for all Miller-Bravais face symbols $h + k + i = 0$.

As an example of work in this notation, we may derive the index of the face at the intersection of the zones $[20\bar{2}1, 01\bar{1}0]$ and $[10\bar{1}0, 01\bar{1}1]$. The cross-multiplications are performed with the pairs of indices $[20*1, 01*0]$ and $[10*0, 01*1]$:

$$
\begin{array}{c|cccc|c}
2 & 0 & 1 & 2 & 0 & 1 \\
  &   & \times & \times & \times & \\
0 & 1 & 0 & 0 & 1 & 0 \\
\hline
\end{array}
\qquad\qquad
\begin{array}{c|cccc|c}
1 & 0 & 0 & 1 & 0 & 0 \\
  &   & \times & \times & \times & \\
0 & 1 & 1 & 0 & 1 & 1 \\
\hline
\end{array}
$$

$$[\bar{1}02] \qquad\qquad\qquad [0\bar{1}1]$$

leading to zone symbols $[\bar{1}0 \dagger 2]$ and $[0\bar{1} \dagger 1]$.

Cross-multiplying these symbols:

$$
\begin{array}{c|cccc|c}
\bar{1} & 0 & 2 & \bar{1} & 0 & 2 \\
  &   & \times & \times & \times & \\
0 & \bar{1} & 1 & 0 & \bar{1} & 1 \\
\hline
\end{array}
$$

$$(211)$$

we derive the face symbol $(21*1)$, and the required index is $(21\bar{3}1)$.

THE SINE RATIO (OR ANHARMONIC RATIO) OF FOUR TAUTOZONAL FACES

If P_1, P_2, P_3, P_4 are four faces in a zone, no two of which are parallel, and we denote by θ_{12} the angle $P_1 \frown P_2$, and so on, then the trigonometrical ratio

$$\frac{\dfrac{\sin \theta_{12}}{\sin \theta_{13}}}{\dfrac{\sin \theta_{42}}{\sin \theta_{43}}}$$

is termed the *sine ratio* of the four tautozonal faces. (It is sometimes termed alternatively the *anharmonic ratio*, but since harmonic ratios are included also as special, and important, cases (p. 213), this name is inappropriate and should not be used.) We first proceed to show by a simplified demonstration that on the basis of our assumptions con-

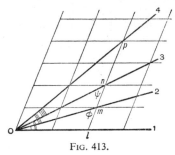

FIG. 413.

cerning the relationship of crystal faces to the underlying structural pattern, this ratio must have some simple value.

In Fig. 413 a portion of a crystal pattern is drawn on a plane normal to a prominent zone-axis, and the directions of four faces in this zone are shown by lines 1, 2, 3, 4 passing through a common point of the pattern.

From plane triangles we have

$$\frac{\sin \theta_{12}}{\sin \theta_{13}} = \frac{lm \sin \phi}{ln \sin \psi}$$

and

$$\frac{\sin \theta_{42}}{\sin \theta_{43}} = \frac{pm \sin \phi}{pn \sin \psi},$$

whence the sine ratio

$$\frac{\dfrac{\sin \theta_{12}}{\sin \theta_{13}}}{\dfrac{\sin \theta_{42}}{\sin \theta_{43}}} = \frac{\dfrac{lm}{ln}}{\dfrac{pm}{pn}}.$$

Since lm, etc., are simply related to a unit of the structural pattern, this ratio clearly has a simple rational value.

To determine this value in terms of the indices of the faces, we proceed as follows. In Fig. 414, P_1, P_2, P_3, P_4 are the poles of the four tautozonal faces $h_1 k_1 l_1$, $h_2 k_2 l_2$, $h_3 k_3 l_3$, $h_4 k_4 l_4$, and x, y the poles of two of the crystallographic axes. Draw the great circles $x P_1$, $y P_1$, etc.; for brevity, we denote the values of these arcs by x_1, y_1, etc.

In $\triangle x P_1 P_2$,

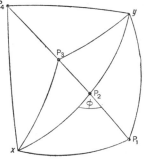

FIG. 414.

$$\cos x_1 = \cos \theta_{12} \cos x_2 + \sin \theta_{12} \sin x_2 \cos \phi. \quad \dots\dots\dots\dots(1)$$

And in $\triangle x P_4 P_2$,

$$\cos x_4 = \cos \theta_{42} \cos x_2 - \sin \theta_{42} \sin x_2 \cos \phi. \quad \dots\dots\dots\dots(2)$$

Multiplying equation (1) by $\sin \theta_{42}$,

and ,, (2) by $\sin \theta_{12}$, and adding:

$$\cos x_1 \sin \theta_{42} + \cos x_4 \sin \theta_{12} = \cos x_2 (\sin \theta_{42} \cos \theta_{12} + \cos \theta_{42} \sin \theta_{12})$$
$$= \cos x_2 \sin \theta_{14}.$$

Similarly, from the $\triangle$'s $y P_1 P_2$ and $y P_4 P_2$:

$$\cos y_1 \sin \theta_{42} + \cos y_4 \sin \theta_{12} = \cos y_2 \sin \theta_{14}.$$

By division,

$$\frac{\cos x_1 \sin \theta_{42} + \cos x_4 \sin \theta_{12}}{\cos y_1 \sin \theta_{42} + \cos y_4 \sin \theta_{12}} = \frac{\cos x_2}{\cos y_2}. \quad \ldots\ldots\ldots\ldots\ldots(3)$$

Now, from the equations of a normal, we can write

$$\frac{a}{h_1} \cos x_1 = \frac{b}{k_1} \cos y_1, \text{ etc.,}$$

and eliminating y_1, y_4 from equation (3):

$$\frac{\cos x_1 \sin \theta_{42} + \cos x_4 \sin \theta_{12}}{\dfrac{a}{b}\dfrac{k_1}{h_1} \cos x_1 \sin \theta_{42} + \dfrac{a}{b}\dfrac{k_4}{h_4} \cos x_4 \sin \theta_{12}} = \frac{\cos x_2}{\dfrac{a}{b}\dfrac{k_2}{h_2} \cos x_2}$$

$$\left(\frac{k_2}{h_2} - \frac{k_1}{h_1}\right) \cos x_1 \sin \theta_{42} = \left(\frac{k_4}{h_4} - \frac{k_2}{h_2}\right) \cos x_4 \sin \theta_{12},$$

$$\frac{\sin \theta_{12}}{\sin \theta_{42}} = \frac{h_1 k_2 - k_1 h_2}{h_2 k_4 - k_2 h_4} \cdot \frac{\cos x_1}{\cos x_4} \cdot \frac{h_4}{h_1}.$$

From consideration of the poles P_1, P_3, P_4 in place of P_1, P_2, P_4, it follows similarly that

$$\frac{\sin \theta_{13}}{\sin \theta_{43}} = \frac{h_1 k_3 - k_1 h_3}{h_3 k_4 - k_3 h_4} \cdot \frac{\cos x_1}{\cos x_4} \cdot \frac{h_4}{h_1},$$

and by division:

$$\frac{\dfrac{\sin \theta_{12}}{\sin \theta_{13}}}{\dfrac{\sin \theta_{42}}{\sin \theta_{43}}} = \frac{\dfrac{h_1 k_2 - k_1 h_2}{h_1 k_3 - k_1 h_3}}{\dfrac{h_2 k_4 - k_2 h_4}{h_3 k_4 - k_3 h_4}}.$$

By similar use of the pole of the axis z, and indices l, etc., it follows that the sine ratio is also equal to similar expressions involving k and l indices, and l and h indices respectively. Moreover, the expressions $(k_1 l_2 - l_1 k_2)$, etc., are the zone symbols U, V, W calculated from the appropriate pairs of indices. Hence we may write finally:

$$\frac{\dfrac{\sin \theta_{12}}{\sin \theta_{13}}}{\dfrac{\sin \theta_{42}}{\sin \theta_{43}}} = \frac{\dfrac{U_{12}}{U_{13}}}{\dfrac{U_{42}}{U_{43}}} = \frac{\dfrac{V_{12}}{V_{13}}}{\dfrac{V_{42}}{V_{43}}} = \frac{\dfrac{W_{12}}{W_{13}}}{\dfrac{W_{42}}{W_{43}}}. \quad \ldots\ldots\ldots\ldots\ldots\ldots(4)$$

THE LAW OF RATIONAL SINE RATIOS (MILLER'S LAW)

The above relationship was first published by W. H. Miller, in his *Treatise on Crystallography*, in 1839, but seems to have been already realised by other crystallographers also. The values of U, V, W on the right-hand sides of the equations, being zone indices, must be whole numbers, and it thus follows that *the sine ratio of four tautozonal faces is a commensurable number.* The *Law of Rational Sine Ratios* thus established is the basis for formal proof of many of the propositions for which we have already assumed results. We can now prove, for example, that a pentad symmetry axis is impossible by showing that four tautozonal faces 72° apart would result in an irrational sine ratio, and by generalising the method of attack we can prove that the *only* possible axes of rotational symmetry are those of degree 2, 3, 4 and 6.

The equations (4) are used in practice to solve two problems of constant recurrence in crystallographic calculations:

(1) Given the angular positions of four faces in a zone, and the indices of three of the faces, to determine the indices of the fourth.

(2) Given the indices of four faces in a zone and the angular relationships of three of them, to determine the position of the fourth.

The method of working is most easily displayed in terms of actual examples. It must be impressed upon the student that orderliness is essential; the poles should be numbered 1, 2, 3, 4 along the zone in question, and the calculation of zone symbols be carried out in strict conformity with the subscripts of the angles. To calculate $[U_{43} \, V_{43} \, W_{43}]$, for example, write down the index 4 twice for cross-multiplication, and the index 3 twice *below* it; reversal of this order would change the sign of the co-ordinates, and the calculation would be incorrect unless the 42 indices were also reversed.

Examples.

(1) In a triclinic crystal, $001 \wedge 111 = 44° \, 29'$; $001 \wedge 112 = 28° \, 55'$. A fourth face occurs in this zone, between 112 and 111, at an angular distance of $35° \, 16'$ from 001; what is its index?

	1	2	3	4							
	001	112	hkl	111		0	0 1 0 0	1			
						1	1 2 1 1	2			
				1, 2			$[\bar{1}10]$				

$$\frac{\begin{array}{l}\sin 28° 55'\\ \sin 35° 16'\\ \hline \sin 15° 34'\\ \sin 9° 13'\end{array}}{} \quad \frac{\bar{1}}{\bar{k}}\ \frac{1}{h}\ \frac{0}{0} = \frac{1}{1}\ \frac{1}{\bar{1}}\ \frac{0}{0} = \frac{0}{l-k}\ \frac{0}{h-l}\ \frac{0}{k-h}$$

$$\tfrac{1}{2} = \frac{l-k}{k} = \frac{l-h}{h}$$

$$3h = 3k = 2l$$

$$\frac{h}{2} = \frac{k}{2} = \frac{l}{3}$$

and the required index is (223).

(2) In another triclinic substance,

$$100\frown 010 = 79° 6' \ ; \quad 100\frown 110 = 26° 7'.$$

What is the angle $100\frown 120$?

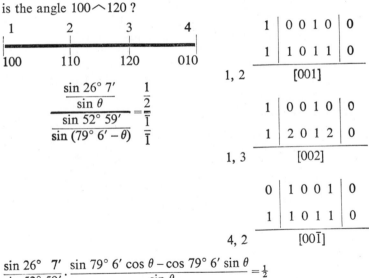

```
1        2        3        4
100     110      120      010
```

$$\frac{\begin{array}{l}\sin 26° 7'\\ \hline \sin \theta\\ \sin 52° 59'\\ \sin (79° 6' - \theta)\end{array}}{} \quad \frac{1}{2}\ \frac{\bar{1}}{\bar{1}}$$

$$\frac{\sin 26°\ 7'}{\sin 52° 59'} \cdot \frac{\sin 79° 6' \cos \theta - \cos 79° 6' \sin \theta}{\sin \theta} = \tfrac{1}{2}$$

Right column determinant tables:

```
          0 | 0 1 0 0 | 1
          h | k l h k | l
1, 3      ------------------
                 [k̄ h 0]

          1 | 1 1 1 1 | 1
          1 | 1 2 1 1 | 2
4, 2      ------------------
                 [1 1̄ 0]

          1 | 1 1 1 1 | 1
          h | k l h k | l
4, 3      ------------------
             [l − k, h − l, k − h]

          1 | 0 0 1 0 | 0
          1 | 1 0 1 1 | 0
1, 2      ------------------
                 [001]

          1 | 0 0 1 0 | 0
          1 | 2 0 1 2 | 0
1, 3      ------------------
                 [002]

          0 | 1 0 0 1 | 0
          1 | 1 0 1 1 | 0
4, 2      ------------------
                 [00 1̄]

          0 | 1 0 0 1 | 0
          1 | 2 0 1 2 | 0
4, 3      ------------------
                 [00 1̄]
```

$$\cot \theta = \tfrac{1}{2} \cdot \frac{\sin 52° \ 59'}{\sin 26° \ 7' \sin 79° \ 6'} + \cot 79° \ 6'$$

$$= 0.9237 + 0.1926$$

$$= 1.116$$

$$\underline{\theta = 41° \ 51'}$$

TRANSFORMATIONS OF THE SINE RATIO

1. *The co-tangent relationship.*

The sine ratio $\dfrac{\dfrac{\sin \theta_{12}}{\sin \theta_{13}}}{\dfrac{\sin \theta_{42}}{\sin \theta_{43}}} = \dfrac{\sin \theta_{12} \cdot \sin (\theta_{14} - \theta_{13})}{\sin \theta_{13} \cdot \sin (\theta_{14} - \theta_{12})}$

$$= \frac{\sin \theta_{12} (\sin \theta_{14} \cos \theta_{13} - \cos \theta_{14} \sin \theta_{13})}{\sin \theta_{13} (\sin \theta_{14} \cos \theta_{12} - \cos \theta_{14} \sin \theta_{12})}$$

$$= \frac{\sin \theta_{12} \sin \theta_{13} \sin \theta_{14} (\cot \theta_{13} - \cot \theta_{14})}{\sin \theta_{12} \sin \theta_{13} \sin \theta_{14} (\cot \theta_{12} - \cot \theta_{14})}$$

$$= \frac{\cot \theta_{13} - \cot \theta_{14}}{\cot \theta_{12} - \cot \theta_{14}},$$

whence, if the numerical value of the ratio $= \dfrac{p}{q}$, we have

$$p \cot \theta_{12} - q \cot \theta_{13} = (p - q) \cot \theta_{14}, \quad \dotsfill (5)$$

a form in which calculations can be readily carried out with the help of a table of co-tangents.

In example (2) above, for instance,

$$\theta_{12} = 26° \ 7', \quad \theta_{14} = 79° \ 6', \quad \frac{p}{q} = \tfrac{1}{2},$$

$$\cot 26° \ 7' - 2 \cot \theta_{13} = - \cot 79° \ 6',$$

$$\cot \theta_{13} = \tfrac{1}{2}(2.0398 + 0.1926)$$

$$= 1.116,$$

whence $\underline{\theta_{13} = 41° \ 51'.}$

2. *The tangent relationship.*

If $\theta_{14} = 90°$ a still further simplification arises, since the sine ratio

$$= \frac{\dfrac{\sin \theta_{12}}{\sin \theta_{13}}}{\dfrac{\cos \theta_{12}}{\cos \theta_{13}}} = \frac{\tan \theta_{12}}{\tan \theta_{13}},$$

whence $\tan \theta_{12} = \dfrac{p}{q} \tan_{13},$

and we can work directly with a table of natural tangents (or more rapidly still, with one of ' multiple tangents '). It is often possible in all the more symmetrical systems so to select P_1 and P_4 for a given calculation that $P_1 \frown P_4 = 90°$, and this important simplification is effected.

THE GRAPHICAL SOLUTION OF SINE RATIOS

If in Fig. 413 the line $l\ m\ n\ p$ were parallel to face 4, we can write

$$\frac{\dfrac{\sin \theta_{12}}{\sin \theta_{13}}}{\dfrac{\sin \theta_{42}}{\sin \theta_{43}}} = \frac{lm}{ln}.$$

Similarly, in Fig. 415, if four face-poles are situated on the primitive zone circle at P_1, P_2, P_3, P_4 and a line $l\ m\ n$ be drawn across their normals parallel to the normal to P_4, then the value of the sine ratio is given by the ratio $\dfrac{lm}{ln}$. This leads to a simple graphical solution of the two problems of p. 210. If the angular relationships of all four faces are known, we can plot them around a circle, draw their normals, draw $l\ m\ n$ across the pencil of normals parallel to OP_4, and read off the ratio $\dfrac{lm}{ln}$. If θ_{13} and θ_{14} are known, and all four

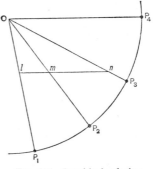

FIG. 415. Graphical solution of sine ratios.

indices, we can mark off along the line $l\ n$ the calculated value of the sine ratio, and thus construct the normal to the face P_2. (Graphical solutions are also possible when other pairs of angles, such as θ_{12} and θ_{34} are known, but such cases are unlikely to arise in practice.)

A special interest of the graphical solution lies in its ability to detect at sight the particular case where $l\ m = m\ n$, and the ratio $= \frac{1}{2}$, the *harmonic ratio* of mathematicians. Harmonic ratios are particularly common as a consequence of the usual simplest relationship of the indices of actual faces (possibly 95% of calculations in crystallography, excluding the province of minerals where complex habits are more frequent); notice that both the examples (p. 210) are harmonic. In such cases, formula (5) (p. 212) simplifies still further:

$$\cot \theta_{12} + \cot \theta_{14} = 2 \cot \theta_{13}.$$

CHAPTER IX

CRYSTAL DRAWINGS

DRAWING CRYSTALS FROM A PROJECTION

The only method of crystal drawing which we have so far discussed has involved the preparation of the appropriate axial cross in clinographic projection. On this cross we have represented faces and edges by consideration of indices as representing sub-multiples of the axial units. The derivation of zone symbols provides another method of construction, particularly useful sometimes if high indices are involved. To determine the direction of the edge between the faces 896 and 98$\bar{6}$, for example, in a pentagonal icositetrahedron {968} (Fig. 296), we need not divide the axial units into eighths and ninths, but instead cross-multiply the indices and plot co-ordinates proportional to the resulting values of U, V, W. Drawing on an axial cross is a valuable exercise for the student, for it brings out very clearly the meaning of indices and the relationships between indices and intercepts; it is the quickest method of representing single forms and simple combinations, but it suffers from grave disadvantages when one is dealing with crystals of more complex development. If the point of view adopted for the standard clinographic axial cross proves unsuitable for giving a clear representation of a particular habit, a change of direction can be effected only by construction of a completely new axial cross. Moreover, in developing a complex habit by successive modification of simpler combinations it is very difficult to control in detail the precise habit of the finished drawing. The practised crystallographer therefore draws all but the simplest combinations directly from a projection (whether stereogram or gnomonogram), in which all the required information concerning inclination of edges is incorporated in the representation of the corresponding zones.

DRAWING FROM THE STEREOGRAPHIC PROJECTION

The first step is the preparation of a plan of the crystal as seen by parallel projection from above—an *orthographic plan* on the plane of stereographic projection. In such a plan all the edges between faces in any one zone will be represented by a series of parallel lines, and the common direction for any one zone is given by the normal to the

diameter of the zone circle in the stereogram. In Fig. 416 is developed
the plan of the crystal of borax of Fig. 125, and the different stroke
used for each zone shows clearly the manner of derivation. It is easy

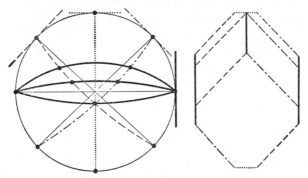

FIG. 416. Construction of an orthographic plan from a stereogram.

to make such a plan as closely representative of a particular habit as
we may wish; allowance must be made for the decrease in apparent
width of faces as they slope more steeply to the vertical axis, but this
decrease is proportional to the cosine of the angle of slope, and can thus
be precisely evaluated. Compare, for example, the plan view of the
complex crystal of sulphur, Fig. 417, with the finished representation
made from it, Fig. 124.

The finished parallel-perspective drawing is obtained by constructing
a new ' plan ' as a projection on a non-crystallographic plane normal

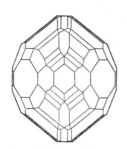

to the chosen direction of view. Since
this, also, is an orthographic plan, the
drawings finally obtained are *orthographic*
representations, and not clinographic drawings like those which we have hitherto constructed on the Naumann axial cross. As
already suggested by Figs. 63 and 64, the
difference in appearance is inappreciable in
practice.

FIG. 417. The orthographic
plan of a sulphur crystal from
which Fig. 124 was constructed.

The next step, then, is to insert in the
stereogram the pole of the direction parallel
to which the crystal is to be viewed, and here
appears the second feature in which this method of drawing is much more
flexible than a construction on an axial cross. The conventional view-
point lies in the right-hand upper front octant at a position *W*, involv-

ing a rotation θ of about $18\frac{1}{2}°$ and an elevation ϕ above the horizontal plane of between 6° and 10° (Fig. 418), the plane of the required projection being the great circle AQB of which W is the pole, but we can easily vary the values of θ and ϕ at will to change the point of view. The great circle AQB is now traced on the stereogram; if this trace should pass through, or very near, the pole of a face on the crystal, such a face will be foreshortened almost into a line in the final drawing; it is then advisable to change the position of W, by suitable

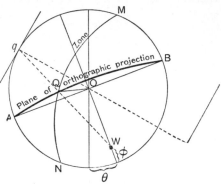

FIG. 418. Construction of the direction of a zone axis in an orthographic drawing.

changes in the values of θ and ϕ, before proceeding further.

To find the direction in the new ' plan ' of edges corresponding to a particular zone we require, in effect, to rotate the projection until W lies in the centre, when the normal to the diameter of the zone in this new position gives the required direction as before. Suppose such an inclined zone NQM cuts the plane of projection AQB at Q. Then OQ

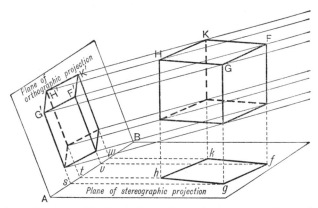

FIG. 419. The relationship of an orthographic projection on an arbitrary plane to an orthographic plan on the plane of stereographic projection.

is the diameter of the zone on the plane of projection AQB, and we know from the properties of the pole of a zone (p. 29) that if Q is projected from W to q on the primitive the arc AQ = arc Aq. Hence the tangent to the primitive at *q* gives the zonal direction which we require in our drawing.

Having obtained the required direction of edge we must now determine its correct length, and this is found from the original orthographic plan on the plane of the primitive. All vertical lines in the crystal will project normal to the diameter *AB*, that is, as parallels to *OW*, and by drawing a series of such parallels from the coigns of the original plan we can limit the lengths of edges in the finished drawing. The relationship of the two ' plans ' is most easily seen by reference to Fig. 419. The coigns *F, G, H, K* of the crystal project as *f, g, h, k* in the plan on the plane of stereographical projection, and as F^1, G^1, H^1, K^1 on the chosen plane of orthographic projection. If now the plane of orthographic projection were revolved about the line *AB* (the diameter *AB* of Fig. 418) to coincide with the plane of stereographic projection, it is clear that F^1, G^1, H^1, K^1 will fall respectively on the lines *fv, gs, ht, kw*, normals to *AB* from *f, g, h, k*.

The simplest way to follow this description is to carry the instrucions out in practice, and the student is advised next to reproduce for

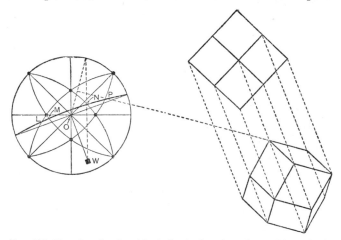

FIG. 420. Drawing the rhombic dodecahedron in orthographic projection.

himself the construction of Fig. 420, illustrating the drawing of a rhombic dodecahedron. The four zone circles intersect the trace of the plane of orthographic projection in the points *L, M, N, P*, which are projected on to the primitive from *W* to give the four directions of edge required. This simple example shows no vertical edges; such edges on a crystal project, of course, as parallels to *OW*, and their length is easily chosen to give a correct representation of the required habit, in proportion to the size of the terminal faces. In centrosymmetrical crystals, the lower half can be completed by drawing parallels, since no new zonal directions are involved; in crystals without a centre of symmetry the faces on the lower half must be correctly represented in the stereogram, but it is convenient to replace all such poles by the pole of the parallel face on the upper hemisphere before beginning the drawing.

DRAWING FROM THE GNOMONIC PROJECTION

Instead of starting from a stereogram we may draw directly from a gnomonogram, which indeed was the type of projection from which this method of drawing was first evolved. An orthographic plan on the plane of gnomonic projection is first prepared, a proceeding precisely analogous with the corresponding step in stereographic procedure. The required directions of edge are found as normals to the corresponding zone-lines. The trace of the intersection of the plane of the final orthographic projection is also drawn (corresponding to the great circle AQB of Fig. 418) at a distance $r \tan \phi$ from the centre and inclined at the chosen angle θ. This trace is known as the *guide-line*.

We now require to find the corresponding *angle-point*, at which any two points on the guide-line subtend in the projection the true angle between the directions which they represent. (The guide-line, of course, is not a zone-line; the angle-point is used in ordinary gnomonic work to measure the interfacial angles between poles on a zone-line.) In Fig. 421 (*a*) the construction is represented in three dimensions. *O* is

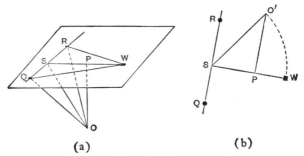

(a)   (b)

FIG. 421. The angle-point construction.

the centre of the sphere, OP the normal to the plane of gnomonic projection; Q and R two points on the guide-line. Then the angle

$Q\hat{O}R$ at the centre of the sphere is the angle we require to measure. Draw SP normal to QR and produce to W; if $SW = SO$ it is clear that $Q\hat{W}R = Q\hat{O}R$. In projection (Fig. 421 (b)) we draw PS normal to QR, and PO' normal to PS and equal to the radius of the sphere of projection. Then W lies on SP produced, at a distance $SW = SO'$.

Having located the angle-point of the guide-line we can proceed in gnomonic projection perhaps even more easily than in stereographic. For any required zonal direction, join the angle-point to the intersection of the zone with the guide-line, and draw the edge in a direction normal to this join. The method is shown in Fig. 422, where it is used to complete a drawing of borax from a plan similar to Fig. 416.

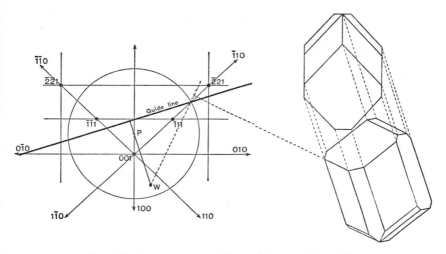

FIG. 422. Drawing a crystal of borax from a gnomonogram.

The special advantage of the gnomonic projection as a basis for drawing lies obviously in the fact that the straight zone-lines and guide-line are more easily drawn than the great circles of the stereogram, some of which may possibly be of very large radius. It is at a disadvantage only if some of the intersections of zone-lines with the guide-line lie at a great distance from the centre of the projection. The neatest method is clearly a combination of both, the gnomonic projection being used for speed and accuracy until a remote intersection arises which is more easily located on the great circle AQB of the stereogram. This changing to and fro offers no inconvenience in practice, for the stereographic pole corresponding to given values of θ and ϕ coincides with the angle-

point of the equivalent guide-line (for which reason we have used the
letter W throughout, from the German *Winkelpunkt*).   In Fig. 423, a

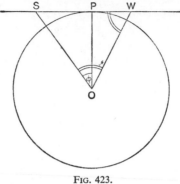

FIG. 423.

vertical section of the sphere through OW, $SO = SW$ from the angle-
point construction;   hence $P\hat{O}W = \dfrac{90° - \phi}{2}$, $PW = OP \tan \dfrac{90° - \phi}{2}$, so
that W is the correct distance from the centre of the projection to
represent stereographically the pole of the plane of orthographic
projection.

PART II

THE SYMMETRY OF THE INTERNAL ARRANGEMENT

CHAPTER X

THE SYMMETRY OF INTERNAL STRUCTURE

THE FOURTEEN BRAVAIS LATTICES AND THE INTERNAL STRUCTURE OF CRYSTALS

So far our direct observations have been confined to the external geometry of the crystal, except when the discussion of such a property as crystal cleavage has brought us to a brief consideration of the interior. The regularities which we have observed, however, and which are summarised in the laws of constancy of angle and of rationality of indices and in the groups of seven crystal systems and thirty-two crystal classes have all combined to convince us that a crystal is an orderly assemblage obtained by the regular repetition of some unit of pattern. The symmetry of this unit determines ultimately the external symmetry of the crystal, and hence the particular system into which it will fall. We no longer suppose with Haüy that these units are *solid* parallelepipeda, and have replaced each unit by a representative point such as its centre of gravity (Fig. 53). We have already used this picture in representing a cubic crystal as an assemblage of skeletal cubelets,

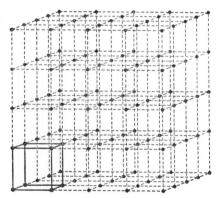

FIG. 424. A portion of a primitive cubic space lattice. A unit of the lattice is heavily outlined.

a tetragonal crystal as one of right square prisms, and so forth. By the repetition of the unit cubelets by parallel translations we have built up a *cubic space lattice* (Fig. 424), an arrangement of points in space, with cubic symmetry, such that the environment of any one point is identical in arrangement and orientation with that of any other point (the array being considered to extend indefinitely). The unit of this space lattice is the simple cube outlined in the figure.

There are, however, other ways of arranging points with identical

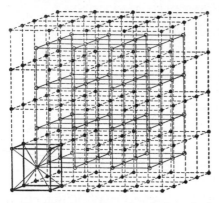

environment in parallel orientation so that the whole array displays the four triad axes characteristic of cubic symmetry. If we start with a unit cube which has a lattice point at the intersection of the body-diagonals as well as at the corners, we obtain by translation the arrangement of Fig. 425, and examination will show that in such an arrangement all the points have identical environments in parallel orientation —it is a further type of cubic space lattice. A third possible arrangement is developed by starting with the unit cube de-

FIG. 425. A portion of a body-centred cubic space lattice. A unit of the lattice is heavily outlined, with an open ring in the body-centring position; throughout the lattice, however, both dots and rings have identical environments.

picted in Fig. 426, with lattice points at the centre of each cube face in addition to the corners. The immediate environment of every point in each of these arrangements is shown in Fig. 427. The simple cube is a *primitive* unit, denoted by the letter P; in the cubic P space lattice every point has six nearest neighbours. The *body-centred* cube unit is

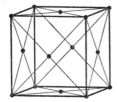

FIG. 426. A face-centred cube.

FIG. 427. The immediate environments of points in the cubic P, I and F space lattices.

denoted by the letter I (German *Innenzentrierte*); in the cubic I space lattice every point has eight nearest neighbours. The *face-centred* cube unit is denoted by the letter F; in the cubic F space lattice every point has twelve nearest neighbours.

In the tetragonal system the simplest unit which we can propose is the right square prism of our earlier discussions, a primitive unit cell which will produce by translation a *tetragonal P space lattice*. A body-centred unit cell also produces by translation a true lattice, a *tetragonal I space lattice*. When we pass to consideration of the possibilities of face-centring, a new feature arises. The 001 (or *C*) faces can be centred without the symmetry demanding also the centring of the prism faces.

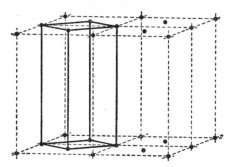

Fig. 428. A primitive tetragonal unit cell (full lines) outlined in a portion of a space lattice built up from a tetragonal *C* face-centred unit (broken lines).

This arrangement proves to be a true lattice, but by an alternative choice of *x* and *y* axes at 45° to their original directions (Fig. 428) the arrangement can be produced by translation of a primitive unit cell of smaller dimensions—a *tetragonal C space lattice* is not a new type of arrangement, but merely equivalent to a tetragonal *P* space lattice in a different orientation. If we next try centring the prism faces alone,

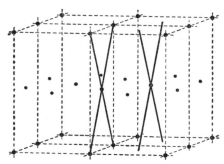

Fig. 429. The full lines show the different environments of two points in a portion of an array built up from a tetragonal unit cell centred on the prism faces.

which the symmetry allows, the arrangement proves not to be a true space lattice, for some points have an environment different from that of others (Fig. 429). Finally, we may centre all the faces, prism and basal pinacoid together, and the student should satisfy himself that translation of such a unit cell produces a true lattice but that it is no new type of arrangement—a *tetragonal F space lattice* is equivalent to a tetragonal I space lattice in a new orientation. Thus there are only two tetragonal space lattices: $P (\equiv C)$ and $I (\equiv F)$.

In the orthorhombic system the simplest arrangement is again a primitive space lattice, orthorhombic P. Centring any one pair of pinacoid faces leads to a true space lattice, but there is a difference of orientation only, and not of type of arrangement, between the products of centring the 001 (C) planes, the 100 (A) planes, or the 010 (B) planes. Conventionally, the orthorhombic one-face-centred arrangement is denoted the *orthorhombic C space lattice*. Body-centred and all-face-centred arrangements are also possible, so that there prove to be four orthorhombic space lattices, P, C, ($\equiv A \equiv B$), I and F (see the chart on p. 226.)

In the monoclinic system we again start with a primitive arrangement. Centring the B (010) faces produces nothing new, for by a

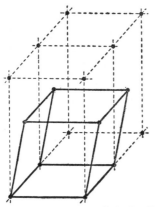

Fig. 430. A monoclinic P cell change in the choice of the direction of
the z axis (Fig. 430) or of the x axis we
can make use of a P cell. Centring the
C (001) faces does produce a new type of
arrangement, a *monoclinic C space lattice*,
and A face-centring is equivalent to this
with a change of orientation. No further
new arrangements are possible—the
student should convince himself that the
monoclinic space lattices F and I can be
described as monoclinic C space lattices
by the appropriate choice of x and z axial
directions. We thus have two different
arrangements in the monoclinic, $P (\equiv B)$,
and $C (\equiv A \equiv F \equiv I)$.

FIG. 430. A monoclinic P cell
outlined in a portion of space
lattice built up by translation of a
monoclinic cell centred on 010
(broken lines).

In the triclinic system only a primitive
unit cell is necessary, since with no restrictions on the choice of axial directions we can always outline a P cell in any triclinic arrangement which is a true space lattice.

In the trigonal system we may choose a rhombohedron as the most appropriate shape of unit cell. Body-centring or face-centring this unit (the symmetry makes all-face-centring the only type of face-centring allowable) produces no new type of arrangement, since we can always choose a new set of rhombohedral edges to outline a primitive cell (Fig. 431). There is thus only one kind of arrangement, which might be denoted as a trigonal *P* space lattice, but it is conventionally given a special symbol, and is described as the *trigonal R space lattice*.

In the hexagonal system, also, only one kind of arrangement proves to be possible, and the unit cell conventionally chosen is a right prism based on a rhombus with an angle of 60° (see chart, p. 226). The best choice of unit cell in trigonal and hexagonal crystals, however, is not always so straightforward as in the other systems, and we shall refer to it more fully later (p. 259), merely noting here that the unit prism of the

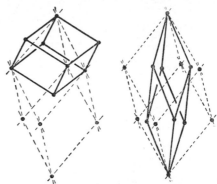

Fig. 431. Primitive rhombohedral unit cells (full lines) outlined in portions of space lattices built up from body-centred and from face-centred units.

hexagonal arrangement was originally denoted *C* not *P*, in spite of its primitive character, and that it sometimes proves to be the most convenient unit to use in describing the internal arrangement of crystals which are morphologically trigonal.

We have thus arrived at fourteen different space lattices:

Cubic   *P, I, F.*
Tetragonal   *P, I.*
Orthorhombic   *P, C, I, F.*
Monoclinic   *P, C.*
Triclinic   *P.*
Trigonal   *R.*
Hexagonal (and trigonal)   *C* (or *P*).

The unit cells, in terms of which we have developed these arrangements, are illustrated together on p. 226. The establishment of the existence of these fourteen different arrangements was the outcome of the work of several mathematicians and crystallographers in the first half of the nineteenth century. Frankenheim * published his first results in 1842, but believed that he had established fifteen different arrangements. Six years later Bravais advanced a more rigid demonstration of the

---

* Moritz Ludwig Frankenheim was born in Brunswick in 1801. After teaching for a short time in the University of Berlin he was appointed to the Chair of Natural Philosophy at Breslau. His crystallographic publications deal chiefly with the elasticity and related physical properties of crystals. He died in Dresden in 1869.

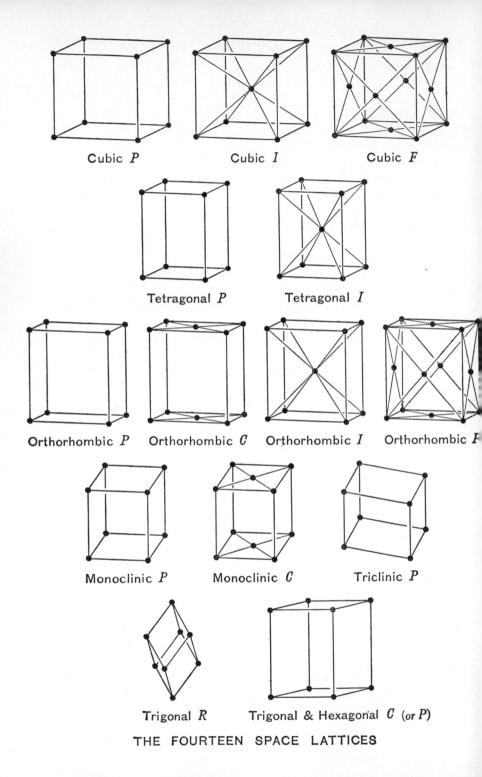

Cubic *P* Cubic *I* Cubic *F*

Tetragonal *P* Tetragonal *I*

Orthorhombic *P* Orthorhombic *C* Orthorhombic *I* Orthorhombic *F*

Monoclinic *P* Monoclinic *C* Triclinic *P*

Trigonal *R* Trigonal & Hexagonal *C* (*or P*)

THE FOURTEEN SPACE LATTICES

possible existence of fourteen different arrangements, pointing out that two of Frankenheim's fifteen were in fact identical. The fourteen space lattices are consequently often called *Bravais lattices*.

The mode of development which we have used differs, of course, from that originally employed by Bravais. It is clearly convenient to describe all the space lattices within one system in terms of a similar set of axes—all cubic lattices in terms of three equal axes at right-angles, all monoclinic lattices in terms of a set of axes in which the x and z directions lie in a plane normal to the diad axis, and so on—just as we did when studying the external morphology. For this reason we have been forced to select in some cases a *multiply primitive unit cell*—one with which more than one equivalent point is associated. In the cubic P cell there is only one equivalent point per unit cell, since each corner is used eight times in the complete assemblage produced by translation. In an I cell, there are two equivalent points per unit cell—the cell is *doubly primitive*—whilst an F cell is *quadruply primitive*. We may note here, however, that it is possible to outline a primitive unit for each of the fourteen arrangements. The primitive unit cell of the cubic F space lattice is the rhombohedron outlined in Fig. 432, a rhombohedron with a plane angle of 60°; the primitive unit cell of

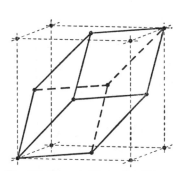

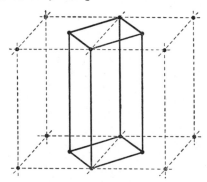

FIG. 432. The primitive unit cell (heavy lines) of a cubic F space lattice.

FIG. 433. The primitive unit cell (heavy lines) of a monoclinic C space lattice.

the monoclinic C space lattice is the oblique rhombic prism of Fig. 433. The student can work out other examples for himself, but it is clear that the frequent change of axial directions within each system necessitated by the use of primitive unit cells only, is highly incon-venient, and we shall be wise in making use of the multiply primitive unit cells already outlined.

The fourteen space lattices all show the full symmetry of the holosymmetric classes of the various systems to which they belong, and Bravais himself realised that the reason for the lower symmetry exhibited by many crystals must lie in the particular arrangement of the structural units (single atoms, groups of atoms, or molecular units) which the identical points of the space lattice represent. The external symmetry elements,—rotation axes, inversion axes and reflection planes —in terms of which we have developed the 32 crystal classes, must all arise from the symmetry of the grouping of the structural units around the lattice points, but perhaps there is more than one kind of internal arrangement which will result in a given external symmetry. If the crystal morphology reveals a vertical diad axis 2, then we may suppose that some structural unit is arranged in pairs about this direction (Fig. 434 *a*); but another kind of two-fold regularity is possible. A rotation through 180° about the diad axis together with a *translation* parallel to the axis will also produce a regularly two-fold arrangement (Fig. 434 *b*). Such an axis is termed a *screw diad axis*, and is denoted 2_1; in diagrams, the screw character is indicated by the ' tails ' seen affixed to the oval

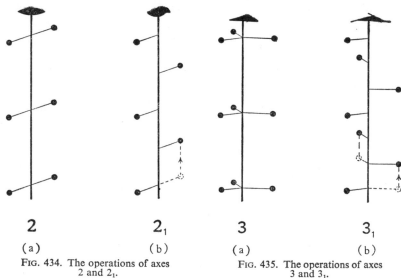

2   2_1 3 3_1

(a)   (b)   (a)   (b)

FIG. 434. The operations of axes   FIG. 435. The operations of axes
2 and 2_1.   3 and 3_1.

flag in the figure. If a morphological diad axis may actually be a screw diad axis in the structure, what are the possibilities with axes of higher degree? Around a direction of three-fold symmetry the structure must show a three-fold arrangement, and the axis may be a true rotation axis 3, normal to plane trigonal groups (Fig. 435 *a*). If, however, every rotation through 120° is combined with a translation parallel to the axis—if the axis is structurally a *screw triad axis*—we may have the arrangement of Fig. 435 *b*. Here each rotation through

120° in an anticlockwise direction is combined with a translation upwards, the fourth position lying in the same vertical plane through the axis as does the first position. Such an axis is conventionally denoted 3_1, but it is not the only kind of screw triad axis possible. If we combine the same translation with the opposite sense of rotation (or translate 2/3 with the same anticlockwise rotation) we have another arrangement around a screw triad axis, 3_2, which is enantiomorphously related to that around 3_1. The two arrangements are illustra-

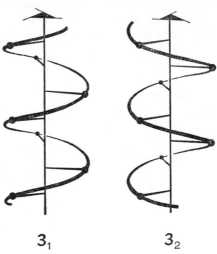

3_1 3_2

FIG. 436. Diagrammatic illustration of the enantiomorphous relationship of axes 3_1 and 3_2.

ted in Fig. 436, with spirals sketched in to help illustrate this relationship. In Fig. 437 they are shown in plan, and it can be clearly seen how the trigonal relationship about the triad axis persists in spite of the different levels indicated by the difference of shading.

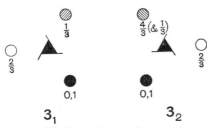

3_1 3_2

FIG. 437. Plans showing the operations of axes 3_1 and 3_2.

Tetrad axes present yet another possibility. Screw tetrad axes 4_1 (translation 1/4 upwards for an anticlockwise rotation of 90°) are enantiomorphously related to screw axes 4_3 (translation 3/4 upwards for the same sense of rotation). We may also have, however, an arrangement about an axis 4_2, where a translation 2/4 is combined with each rotation of 90°, and such an axis has no definite sense of screw, for the same arrangement is reached whether we consider the rotation to be anticlockwise or clockwise. The three arrangements are illustrated in plan in Fig. 438; note the particular devices which display the character of the axis.

The possibilities arising when a rotation hexad axis is revealed by the external morphology should now be clear. There may be true rotation axes 6 in this direction in the internal structure also; but the details of the internal arrangement in some crystals may be consistent only with screw hexad axes. For an axis 6_1, the translation is 1/6 upwards for a rotation anticlockwise through 60°, whilst 6_5 is the symbol

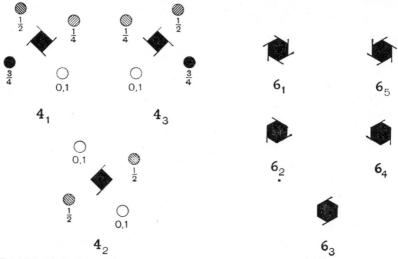

Fig. 438. Plans showing the operations of the three kinds of screw tetrad axes.

Fig. 439. The indicating devices used to show the five kinds of screw hexad axes.

of the enantiomorphous arrangement. 6_2 denotes a translation of 2/6, and 6_4 the enantiomorphous arrangement. 6_3, with a translation 3/6, has, like 4_2, no definite sense of screw. The indicating devices for these five kinds of screw hexad axes are shown against the corresponding symbols in Fig. 439.

If a rotation axis in the morphological symmetry of a crystal may be the representative of screw axes in this direction in the internal structural arrangement, what can we deduce from the detection of an external plane of reflection symmetry? We shall expect to find planes in the structure in this direction on either side of which structural units are situated with an enantiomorphous relationship. If the external symmetry plane m is also a true reflection plane in the structure the arrangement is that illustrated in two dimensions in Fig. 440 a, where the figure sevens represent some particular structural unit. But the arrangement of Fig. 440 b, of the kind familiar to us all in consequence

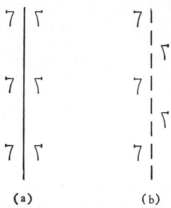

(a)   (b)

FIG. 440. Illustrations of mirror reflection and of glide reflection.

of its popularity as a basis for wallpaper designs, is a closely related one. The reflection operation, however, has been combined with a translation in the plane of the diagram; the vertical plane indicated in cross-section by the heavy dashed line is a *glide reflection* plane. The difference between these two types of internal arrangement would not be appreciable externally, for either would give rise to external crystal planes symmetrically disposed about the plane of reflection.

Where more than one plane of symmetry is revealed externally, any or all may correspond to a glide reflection internally. We illustrate this for the present in two dimensions. Fig. 441 represents a pattern of sevens obtained by the repetition, about the representative points of a rectangular net, of a group of four sevens showing two reflection planes

FIG. 441. A pattern showing two sets of mirror reflections.

at right-angles—one of the elements of design in the border ornament of the ' choice sporting neckerchief ' designed by Ruskin's friend.*   In

FIG. 442. In this pattern there is one set of mirror reflections and one set of glide reflections.

Fig. 442 the pattern shows true reflection planes in one direction only, and glide reflections in the direction at right-angles, whilst in Fig. 443 the pattern is founded wholly on glide reflections.

FIG. 443. A pattern based on glide reflections only.

Thus it appears that the class of symmetry to which a crystal belongs is determined by the symmetry of grouping of the structural units which we associate with the points of the underlying Bravais lattice. If we place around every point of a monoclinic space lattice (whether

* *The Two Paths*, Lect. III, ' Modern Manufacture and Design '.

monoclinic *P* or monoclinic *C*) a group which itself possesses a plane of symmetry and a diad axis, then the resultant structure will build up a crystal belonging to the holosymmetric monoclinic class $2/m$; but if the units themselves show only a plane of symmetry, or only a diad axis, then the structure will be one appropriate to class *m* or class 2 respectively. Moreover, many different types of internal arrangement will result in the same external symmetry. In class *m*, for example, the underlying lattice may be *P* or *C*, and the plane of symmetry may be represented internally, in either case, by true reflections or by glide reflections. In class 2, whether the underlying lattice be *P* or *C*, there may be true rotational diad axes parallel to the *y* direction in the structure, or there may be screw diad axes, whilst in the holosymmetric class $2/m$ we must consider possible combinations of rotation diads or screw diads with reflection planes or glide planes and base the patterns on either of the appropriate space lattices—there are several (in some cases many) possible types of structural symmetry *isomorphous* * with each of the 32 crystal classes.

These ideas developed gradually in the half-century following the work of Bravais. L. Sohncke,† adapting to crystallography some results obtained by pure mathematicians, was the first to investigate the result of admitting screw axes. In a Bravais space lattice the environment of every point is identical with that of every other point and is similarly orientated; but Sohncke showed that if identity of environment only, but without necessarily similar orientation, be required, there are 65 such regular *point systems*. (In his first account, published in 1879, he arrived at a total of 66 but later showed that two of these were identical.) In this way he was able to suggest possible types of structural arrangement which would confer on crystals the symmetry of many of the lower classes, but others, notably those which we have described as hemimorphic, were still unaccounted for.

The final development of considering the introduction also of the operations of reflection and inversion marks ' one of the most remark-

---

* This is the term customarily used by mathematical crystallographers to describe the relationship of a group of symmetry elements (a space group, p. 235) to the crystal class (or point group, p. 235) to which it belongs. The same word is also used in chemical crystallography to describe the relationship between two or more substances of related chemical constitution and similar morphological development, but slightly different values of corresponding interfacial angles.

† Leonhard Sohncke was born in 1842; he published his first important crystallographic work, *Entwickelung einer Theorie der Krystallstruktur*, in 1879 whilst Professor of Physics at Karlsruhe. He later held posts at Jena and at Munich and produced important further contributions to crystallography. He died in 1897.

able instances of independent discovery on record '. E. S. Fedorov *
began to publish his results in 1885, and completed them by 1890, but
as he wrote in Russian his work was not at first noticed by crystallographers of other nationalities. A. M. Schoenflies † published his
work in German in 1891, and three years later a British scientist,
W. Barlow,‡ also independently announced the conclusion of his own
investigations. Though they approached the problem from different
points of view, all three arrived at the same result, that these new
operations admit the possibility of 165 further types of arrangement, a
total of 230 in all. In this way all the 32 classes were accounted for,
and in the next chapter we shall investigate the nature of the groups
of symmetry elements which underlie these different kinds of arrangement.

* Evgraf Stepanovich Fedorov, a celebrated Russian crystallographer and mineralogist, was born at Orenburg in 1853. After a brief career in the army he took up
the study of geology, and at the time of publishing his important contributions to
the theory of crystal structure he was in charge of the collections of the Russian
Geological Survey. In 1905 he became Professor of Crystallography and Mineralogy
at St Petersburg, a post which he held for the rest of his life. He devised many
special instruments to help him in his investigations of crystals, including a two-
circle goniometer and a universal microscope-stage. At the time of his death in
1919 he had completed the formidable task of classifying and tabulating all the
available data relating to the morphology of crystalline substances, and had evolved
a method of crystallochemical analysis by which any substance in the tables could
be identified by goniometric measurements. *Das Krystallreich*; *Tabellen zur
krystallo-chemischen Analyse* was published posthumously (written in German) as
a memoir of the Russian Academy of Sciences in 1920.

† Arthur Moritz Schoenflies was born at Landsberg in 1853. He published his
book *Krystallsysteme und Krystallstructur*, describing the 230 different arrangements,
in 1891 whilst Privat-dozent in Mathematics at Göttingen University. In 1911 he
became Professor of Mathematics at Frankfurt am Main, where he died in 1928.

‡ William Barlow was born at Islington in 1845, and his career presents an
interesting contrast with those of the professional crystallographer Fedorov and the
mathematician Schoenflies, for he was a London business man who soon acquired
independent means and leisure to devote to studies which attracted him. His first
paper on the internal symmetry of crystals appeared in 1883, eleven years before he
published his derivation of the 230 possible kinds of arrangement. He died at
Stanmore, Middlesex, in 1934.

CHAPTER XI

SPACE GROUPS

THE 230 SPACE GROUPS

The work of Sohncke, Fedorov, Schoenflies and Barlow established the number of different *kinds* of arrangement possible in crystal structures. The *actual* number of different arrangements possible is, of course, infinite; in our two-dimensional pattern of figure sevens (Fig. 441) we might have grouped the sevens more widely or more closely, or have chosen other figures as representing other kinds of structural groups, but the pattern would still show the two sets of planes of symmetry evident in Fig. 441. What we now require to work out is the 230 different kinds of symmetry scaffolding on which such patterns, in the three-dimensional crystal, may be based. Such an arrangement of symmetry elements is called a *space group*; though we shall study the 230 space groups chiefly by means of patterns of points or of geometrical units such as triangles, it is important to grasp at the outset that it is not these units themselves, but the elements of symmetry of their arrangement, which constitute the space group.

The procedure which we shall generally follow will be to work out for each crystal class in turn all the space groups isomorphous with that class. This we shall do by associating with every point of the appropriate Bravais space lattices the elements of symmetry indicated by the crystal class, taking into account the possibilities of rotation axes in the external symmetry being represented by screw axes in the space group, and of reflection planes in the external symmetry being represented by glide planes in the space group. Whereas the symmetry elements of a space group extend through space (the group being regarded as extending indefinitely), the symmetry elements of a crystal class can all be regarded as passing through a single point, the origin of our crystallographic axes, which point is thus not repeated by the symmetry operations. In the study of space groups, therefore, the 32 groups of symmetry which we have established are termed *point groups*.

TRICLINIC SPACE GROUPS

Point group 1.

The only Bravais lattice in question is the primitive triclinic space lattice P. No symmetry elements other than an identity axis are to be

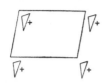

FIG. 444. A portion of a pattern based on the space group $P1$.

associated with each point of this lattice, so that the symbol of the one possible space group is $P1$. Fig. 444 represents diagrammatically in plan a portion of the structure of a crystal based on this space group; a unit of the structure (represented conventionally as a scalene triangle) at some height above the base of the unit cell, indicated by the symbol $+$, is repeated throughout the structure only by the translation of the unit cell.

Point group $\bar{1}$.

Here again only the triclinic space lattice P need be considered, but at each of the lattice points is placed a centre of symmetry (the equivalent of an axis $\bar{1}$). An atomic group associated with the unit of pattern is inverted across a centre to give an enantiomorphous atomic group at an equal distance $(-)$ below the plane of the diagram, as indicated by a shaded triangle (Fig. 445), and translation of this unit

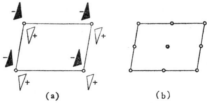

(a)   (b)

FIG. 445. (*a*) A portion of a pattern based on the space group $P\bar{1}$.
(*b*) Plan of a unit of the space group $P\bar{1}$.

produces the pattern, a portion of which is shown in the figure. In Fig. 445 *a* only the centres of symmetry at the corners of the unit of pattern are shown (by thick small circles), but it is clear from the arrangement that other centres of symmetry have arisen, and a plan of a unit of the complete space group $P\bar{1}$ is shown in Fig. 445 *b*.

Thus in the triclinic system there are only two space groups, $P1$ and $P\bar{1}$, but the possibility of describing the structure of an actual triclinic crystal in terms of one of these symbols depends on the correct choice of the crystallographic x, y and z directions to correspond to a primitive unit cell. If these directions have been chosen by reference to the external morphology only, they may correspond to a crystal orientation based on a multiply primitive unit; symbols such as $C\bar{1}$, $F\bar{1}$ and $I\bar{1}$ thus denote the same space group as $P\bar{1}$, but with a different orientation.

MONOCLINIC SPACE GROUPS

Point group 2.

The underlying space lattice may here be either monoclinic P or monoclinic C, and the diad axis of the point group may be parallel to rotation axes or to screw axes in the isomorphous space groups, so that we must consider the possibilities represented by the symbols $P2$,

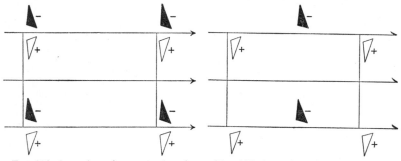

FIG. 446. A portion of a pattern based on the space group $P2$.

FIG. 447. A portion of a pattern based on the space group $P2_1$.

$P2_1$, $C2$ and $C2_1$. In Fig. 446 the representative triangles are repeated in accordance with the presence of diad axes 2 along the y edges of the unit cell (rotation diad axes are represented in plan by double-barbed arrows, headed at one end only if they are uniterminal, as here, where the obtuse corners of all the triangles point towards the left of the figure). The diagram shows clearly that in the space group $P2$ further diad axes arise half-way along the cell in the x direction.

Fig. 447 illustrates similarly the arrangement based on the space group $P2_1$, the horizontal screw diad axes in the y direction being drawn as single-barbed arrows; here again further screw diad axes arise, in addition to those along the edges of the unit cell.

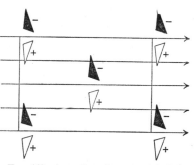

FIG. 448. A portion of a pattern based on the space group $C2$.

If a pattern is based on a monoclinic C space lattice, any element of the pattern associated with the origin 000 of the unit cell is associated also with the point $\frac{1}{2}\frac{1}{2}0$. In Fig. 448 a portion of a pattern based on the space group $C2$ is illustrated; a triangle at a height $+$ is turned

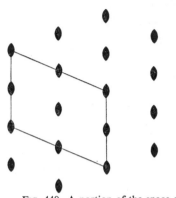

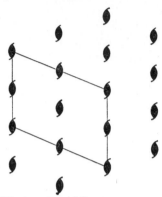

FIG. 449. A portion of the space group $P2$, viewed along the y-axis.

FIG. 450. A portion of the space group $P2_1$, viewed along the y-axis.

over by the operation of a diad axis to one at height $-$, this pair of triangles is repeated also about the C-face-centring point, and the whole pattern is obtained by the appropriate translations. Further diad axes have arisen, as in Fig. 446, but a new feature is evident in the appearance also of screw diad axes between the rotation diad axes. This appearance of screw axes parallel to a direction indicated as one of rotation axes in the space group symbol is a direct consequence of our use of a multiply primitive unit cell. It involves the important deduction here that our suggested space group $C2_1$ will not be a new arrangement, for if we associate screw diad axes with the y edges of the unit of a monoclinic C space lattice, rotation diad axes will also arise in this direction, and the pattern will again be based on the same symmetry as that of Fig. 448, with the arbitrarily chosen origin displaced $\frac{1}{4}$ in the x direction. There are thus only three possible arrangements for an infinite group of parallel diad axes. Portions of the three space groups $P2$, $P2_1$, and $C2$, viewed along the y direction, are illustrated in Figs. 449, 450 and 451 respectively.

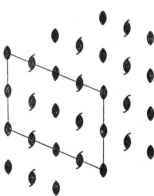

FIG. 451. A portion of the space group $C2$, viewed along the y-axis.

Though there are only three distinct space groups isomorphous with the point group 2, the descriptions which we have given depend on the choice of a particular unit cell. Bearing in mind what was said above (p. 224) concerning the number of different space lattices in the mono-clinic system, it should be clear that a different orientation may lead, for example, to the symbol $B2_1$ in place of $P2_1$, or to a symbol $A2$, $F2$ or $I2$ for the space group conventionally denoted $C2$.

Point group m.

In this point group the space lattices monoclinic P and monoclinic C are to be associated with planes of symmetry normal to the y crystallo-graphic direction, but the mirror reflection plane m of the external symmetry may be represented in the isomorphous space groups either by true reflection planes or by planes of glide reflection. A glide plane is given a symbol denoting the direction of the glide component; this will be an important crystallographic direction, and in a conventional description of these space groups it is always chosen as the direction of the crystallographic z axis and thus is denoted c. The possible space groups which we must consider are thus Pm, Pc, Cm and Cc.

A portion of a pattern based on the space group Pm is illustrated in Fig. 452. Vertical reflection planes parallel to the B faces of the unit

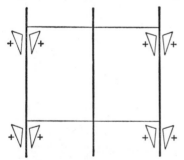

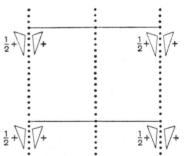

FIG. 452. A portion of a pattern based on the space group *Pm*.

FIG. 453. A portion of a pattern based on the space group *Pc*.

cell are denoted by the thickened lines, and it will be seen that in addition to those reflection planes coinciding with the faces of the unit cell, further true m planes arise half-way between them. In Fig. 453, based on the space group Pc, the vertical planes are c glide planes, indicated in cross-section by dotted lines; a structural unit at a height $+$ above the base of the unit cell is raised to a height $\frac{1}{2}+$ by the opera-tion of glide reflection, and further c glide planes arise in a similar manner to the extra m planes of the space group Pm.

Fig. 454 is a portion of an arrangement based on the space group *Cm*, and is derived from Fig. 452 by associating with the *C*-face-centring point $\frac{1}{2}\frac{1}{2}0$ a group of structural units identical with the group associated with the origin 000. Remembering that we are using a doubly primitive unit in this description, we might be prepared to find that extra symmetry elements arise automatically—planes such as the one marked *ab* in Fig. 454 are planes of glide reflection. The structural unit 1 would reflect in the plane *ab* to the position shown

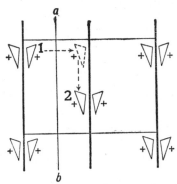

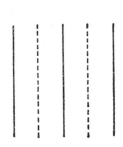

FIG. 454. A portion of a pattern based on the space group *Cm*.

FIG. 455. A portion of the space group *Cm*.

dotted, and a translation *a*/2 would bring it to the position marked 2 —the plane *ab* is an *a* glide plane. Such planes in cross-section are drawn as thick dashed lines, and Fig. 455 illustrates a portion of the space group *Cm* as a series of true reflection planes *m* interleaved by *a* glide planes.

Study of the space group *Cc* should now present little difficulty, for here equally we shall expect further symmetry planes to arise parallel to the *c* glide planes shown in the symbol. In the pattern of Fig. 456 they are seen to be yet a further kind of glide plane, for the translation by which the unit in position 1 is brought to position 2 after reflection involves components *a*/2 and *c*/2—the new planes are *diagonal glide planes* for which we shall later use the symbol *n*. They are indicated in diagrams by the appropriate combination of dot and dash to help remind us of the two

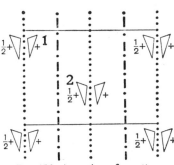

FIG. 456. A portion of a pattern based on the space group *Cc*.

components involved in the glide. (No new group *Pn* would have arisen if we had recognised earlier the possibility of existence of this further type of glide plane, for by the appropriate choice of z axis we can describe *Pn* as *Pc*.)

We have thus established the existence of four space groups, *Pm*, *Pc*, *Cm* and *Cc*, isomorphous with the point group *m*. As before, these particular symbols imply the selection of a particular orientation which may not always agree with an initial choice based on external morphology, so that a number of synonymous symbols can be derived corresponding to other orientations. *Bm*, for example, represents the same space group as *Pm*; *Pa* and *Ba* correspond to other choices of the x and z crystallographic axes in the space group *Pc*, whilst the space groups *Cm* and *Cc* might be described in terms of the monoclinic space lattices *A*, *F* or *I*, each of which is equivalent to the monoclinic *C* space lattice.

Point group 2/m.

The possibilities within this point group are now fairly evident. In the isomorphous space groups based on a primitive space lattice the y direction may be parallel to rotation diad axes or to screw diad axes, whilst the *m* planes of the point group may correspond to true reflection planes or to glide planes in the space group. $P2/m$, $P2_1/m$, $P2/c$ and $P2_1/c$ are therefore four different space groups. When the underlying space lattice is *C*-face-centred, however, screw diad axes will automatically arise parallel to rotation diad axes in the y direction (compare p. 238), so that the only new arrangements here are $C2/m$ and $C2/c$. Moreover, in $C2/m$ there will be *a* glide planes parallel to the *m* planes of the space group, and in $C2/c$ the *c* glide planes will be interleaved by *n* glide planes.

This much should be clear from the preceding study of monoclinic space groups. A further important point arises from consideration of the particular point group now in question. The combination of a diad axis normal to a plane of symmetry resulted automatically in the production of a centre of symmetry in this crystal class; so also in all the space groups isomorphous with the point group $2/m$ centres of symmetry will automatically arise, though it is no more necessary to indicate them in the symbols of the space groups than it was to do so in the point group symbol.

We may illustrate some of these features by working out the space group $C2/m$. Fig. 457 is part of a pattern built up on a *C*-face-centred lattice by the operations of reflection in m planes parallel to 010 combined with rotation about diad axes parallel to crystallographic y. Only these elements of symmetry are inserted in the figure, but the nature of the pattern reveals the presence of further symmetry elements;

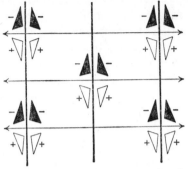

FIG. 457. A portion of a pattern based on the space group $C2/m$.

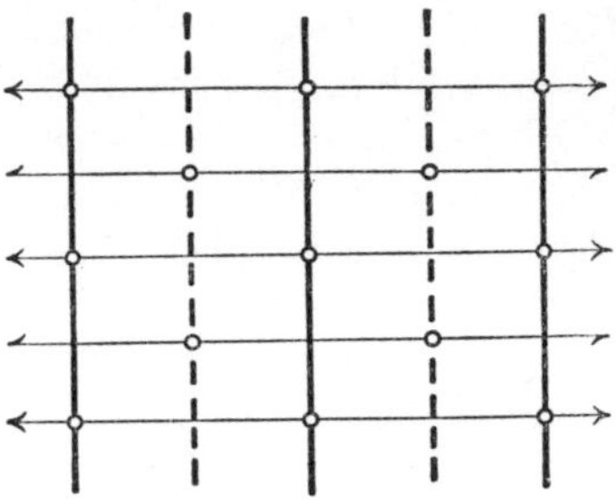

FIG. 458. A portion of the space group $C2/m$.

there are screw diad axes between the rotation diad axes, a glide planes between the m planes, and centres of symmetry at the points of intersection of the rotation diad axes with the m planes and of the screw diad axes with the a planes respectively. A portion of the space group $C2/m$ is shown in Fig. 458.

It should scarcely be necessary to point out once more here that the particular symbols we have derived imply the selection of the most appropriate orientation, and that many synonyms can be derived corresponding to other orientations. We may conclude our examination of monoclinic space groups by tabulating the symbols which we have employed.

Point group 2		Point group m		Point group $2/m$	
$P2$ $P2_1$	$C2$	Pm Pc	Cm Cc	$P2/m$ $P2_1/m$ $P2/c$ $P2_1/c$	$C2/m$ $C2/c$

ORTHORHOMBIC SPACE GROUPS

Point group mm.

The space groups isomorphous with this point group may be based on any one of the four types of orthorhombic space lattice. Considering first those based on the primitive orthorhombic *P* space lattice, the planes parallel to 100 and 010, respectively indicated by the first and second letter *m* of the point group symbol, may be represented in the space groups either as true reflection planes or as glide planes, and we must try all possible combinations. Postulating first that the 100 planes are *m* planes, those parallel to 010 may be *m* planes, *a* planes, *c* planes or *n* planes, and we have the four space groups *Pmm*, *Pma*, *Pmc* and *Pmn*. If the 100 planes are *c* glide planes, we may have the groups *Pca*, *Pcc* and *Pcn*. (The combination represented by *Pcm* is, of course, the same as *Pmc*, but in a new orientation with the *x* and *y* directions interchanged.) Since the *z* direction is uniquely determined as the direction of the intersections of the two sets of vertical planes, we must distinguish also the cases in which the 100 planes are glide planes involving a horizontal component, and we thus derive further new space groups *Pba* and *Pbn*. Finally, the 100 planes may be diagonal glide planes *n*, but the only new combination arising here gives one further space group, *Pnn*. Thus there are in all ten space groups based on a primitive space lattice isomorphous with the point group *mm*.

Before proceeding further we may illustrate some features which arise at this stage by drawing out a pattern based on one of these space groups. Fig. 459 is based on the symmetry elements of the space group which we have derived as *Pbn*. In the figure, however, there are diagonal glide planes *n* parallel to 100, and *a* glide planes parallel to 010; *Pna* is the same space group as *Pbn*, but in an orientation corresponding to the interchange of the *x* and *y* directions. In the positions shown dotted in the figure screw diad axes can be detected, and this should be no surprise to us. Since the point group *mm* involves a diad axis in

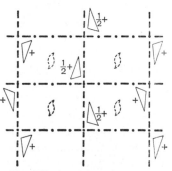

FIG. 459. A portion of a pattern based on the space group *Pna*.

the *z* direction, *all* space groups isomorphous with it must show diad axes in this direction; they may be true rotation diads or screw diads (as in the example figured). The student should work out for himself at this stage patterns based on some of the further space groups we have derived and convince himself that these diad axes do always arise.

Passing next to space groups based on a one-face-centred space lattice, we consider first those based on an orthorhombic *C* lattice.

We know already that, as a consequence of the doubly primitive character of the orthogonal unit cell which we are using, further planes of symmetry arise automatically parallel to any which we may insert in developing a particular space group. Thus the space group *Cmm* will contain glide planes (with a horizontal component of glide) parallel both to 100 and to 010 (*b* planes and *a* planes respectively). Hence such symbols as *Cma*, *Cbm* and *Cba* do not indicate any new combination of symmetry elements and are not used in practice; the conventional symbol for a given space group shows the *highest* symmetry element in a given direction, true reflection planes *m* taking precedence over glide planes *a*, *b* or *c*, and these glide planes taking precedence over diagonal glide planes *n*. *Cmc* does represent a new combination, and we know that in this space group there will be *b* glide planes parallel to 100, interleaved between the *m* planes, and *n* glide planes parallel to 010, interleaved between the *c* planes. Finally, we may have a space group *Ccc* in which no true reflection planes are present, but *c* planes and *n* planes alternate parallel both to 100 and to 010. In all three space groups *Cmm*, *Cmc* and *Ccc* vertical diad axes of some kind are of course present. There are still other space groups based on a one-face-centred space lattice to be considered, for in this crystal class the pinacoid *C*{001} normal to the planes of symmetry is of different significance from the pinacoids *A*{100} and *B*{010} parallel to the planes of symmetry. We must next work out the possible arrangements based on an *A*-face-centred space lattice; a change of orientation of these will correspond to descriptions in terms of *B*-face-centring. As in the preceding group, we are working in terms of a doubly primitive unit cell, so that extra elements of symmetry will arise. *Amm* and *Ama* are new space groups, which can be denoted *Bmm* and *Bbm* respectively if the *x* and *y* directions are interchanged. *Amc* and *Amn*, however, are not new arrangements, for there are *c* planes in *Amm* parallel to 010 and *n* planes in *Ama* parallel to the *a* glide planes. *Abm* is also a new arrangement, for we must distinguish a horizontal glide parallel to the centred face from a horizontal glide parallel to the uncentred 010 planes. Finally, *Aba* is a further new arrangement, in which no true reflection planes exist. As an exercise the student may show that, in all four of these space groups, both rotation diad axes and screw diad axes are present in the *z* direction.

In the body-centred *I* space lattice the point $\frac{1}{2} \frac{1}{2} \frac{1}{2}$ is equivalent to the origin 0 0 0, so that diagonal glide planes *n* will arise parallel to true reflection planes *m*, and *c* glide planes parallel to *a* and *b* planes.

The space group *Imm* is thus the same as the arrangements which might be denoted *Imn* and *Inn*. *Ima* is a new space group, but such arrangements as *Imc*, *Ina* and *Inc* (and their synonyms *Icm*, *Ibn* and *Icn* corresponding to a change of orientation) are equivalent to it. Lastly, *Iba* is new, but contains also *c* glide planes parallel both to 100 and to 010 —*Ica*, *Ibc* and *Icc* are no new arrangements.

The quadruply primitive character of the unit cell of the ortho-rhombic *F* space lattice and the similar significance of the centred *A* and *B* faces restrict even more the number of new arrangements which we can build up on this lattice. Only those arrangements can exist in which the 100 planes and 010 planes are of the same kind. The space group *Fmm* possesses glide planes, which are at the same time *c* and *a* (or *b*) planes, parallel to the true reflection planes, and these re-

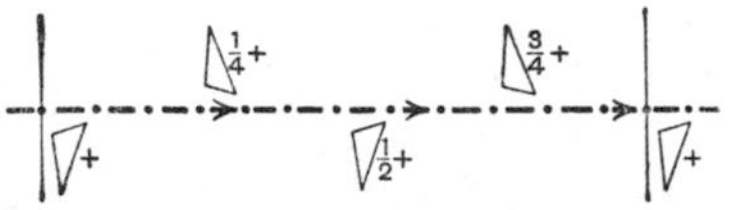

FIG. 460. Diagram to show the operation of a *d* plane.

flection planes are also *n* planes; the only further new space group possible is one in which diagonal glide planes only are present in both directions. These are of a kind which we have not so far encountered, for the glide components are only one-quarter of the primitive translations. For these the symbol *d* is used, and the planes are indicated in cross-section by the dot-dash line of an *n* glide plane to which are added arrows showing the direction in which the glide component in a positive direction is ¼ (Fig. 460). The space group *Fdd* completes our list of twenty-two space groups isomorphous with the point group *mm*, and we tabulate these below before proceeding further.

Primitive P	One-face-centred		Body-centred I	All-face-centred F
	C	A		
Pmm	*Cmm*	*Amm*	*Imm*	*Fmm*
Pma		*Ama Abm*	*Ima*	
Pmc	*Cmc*			
Pmn				
Pca				
Pcc	*Ccc*			
Pcn				
Pba		*Aba*	*Iba*	
Pbn				
Pnn				*Fdd*

Point group 222.

In space groups, isomorphous with this point group, which are based on a primitive space lattice, there may be rotation diad axes in all three

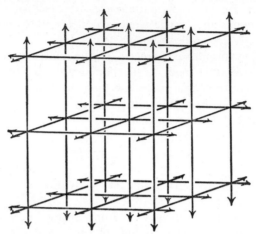

FIG. 461. A portion of the space group *P*222.

directions, in two directions only, in one or in none, screw diad axes of course being present parallel to any of the directions *x*, *y*, *z* not

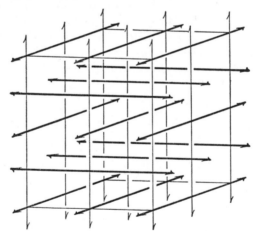

FIG. 462. A portion of the space group $P222_1$.

paralleled by rotation diad axes. Thus $P222$, $P222_1$, $P2_12_12$ and $P2_12_12_1$ are four new space groups, portions of which are illustrated in Figs. 461–464 respectively. Consistently with our usage when studying the crystal morphology, the three figures of the symbol show the character of the x, y and z directions in that order. If the setting of a particular orthorhombic sphenoidal substance is already determined

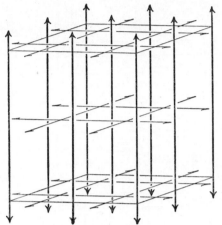

FIG. 463. A portion of the space group $P2_12_12$.

by custom, we may need to write $P22_12_1$ or $P2_122_1$, for example, to describe the space group $P2_12_12$ in a different orientation.

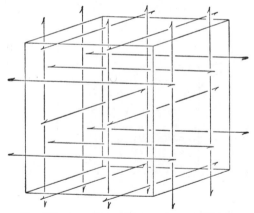

FIG. 464. A portion of the space group $P2_12_12_1$.

The equivalence of the points 0 0 0 and $\frac{1}{2}\frac{1}{2}0$ in the unit cell of an orthorhombic C space lattice necessarily involves, as we know (p. 238), the presence of screw diad axes parallel to any rotation diad axes postulated in the [100] or [010] directions. In the space group $C222$ there are both rotation diad axes and screw diad axes in the x and y

directions but only rotation diad axes in the z direction. The only other possible arrangement based on an orthorhombic C lattice is $C222_1$, in which the z direction shows screw diad axes only. Change of

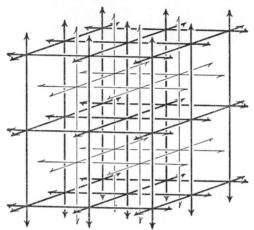

FIG. 465. A portion of the space group $I222$.

orientation would give $A2_122$ or $B22_12$ as synonyms; these symbols do not represent new space groups, since in the crystal class 222 the pinacoids {100}, {010} and {001} are all of similar significance.

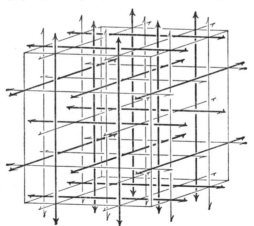

FIG. 466. A portion of the space group $I2_12_12_1$.

The translations involved in an I space lattice result in the space group $I222$ possessing screw diad axes, parallel to the rotation diad axes, in all three axial directions. We might well conclude that no other arrangement based on an I lattice is possible, but in fact there is a second way of arranging screw diad axes parallel to rotation diad axes. Instead of a set of mutually intersecting axes 2 parallel to a set of mutually intersecting axes 2_1 as in $I222$ (Fig. 465), there can also exist a space group in which non-intersecting diad axes 2 are parallel to non-intersecting screw diad axes 2_1 (Fig. 466). To derive a suitable symbol for this second arrangement, we must countenance a departure from the convention that a figure 2_1 in the symbol of a space group indicates screw diad axes only, and no rotation diad axes, in the corresponding direction. The arrangement of axes in Fig. 466 is that of the space group denoted $I2_12_12_1$ (by analogy with the non-intersecting screw diad axes of the group $P2_12_12_1$), in spite of the existence of rotation diad axes in all three principal directions.

Lastly, there exists the space group $F222$, in which screw diad axes also are present in all three directions. We have derived nine space groups isomorphous with the point group 222.

P	C	I	F
$P222$	$C222$	$I222$	$F222$
$P222_1$	$C222_1$		
$P2_12_12$			
$P2_12_12_1$		$I2_12_12_1$	

Point group mmm.

In the space groups isomorphous with this point group and based on a primitive space lattice, planes parallel to 100 may be m planes, b planes, c planes or n planes; those parallel to 010 may be m, a, c or n, and those parallel to 001 m, a, b or n planes. (Notice that the same letter, n, can be used for a diagonal glide plane whatever its orientation. A letter n in the first place in the symbol denotes diagonal glide planes parallel to 100, and the glide must thus be $\frac{1}{2}(b+c)$; similarly for n planes parallel to 010 the glide must be $\frac{1}{2}(c+a)$, and for those parallel to 001 it must be $\frac{1}{2}(a+b)$.) Sixty-four different symbols can be derived from a combination of these planes in sets of three; they do not all represent different space groups, but in some instances merely the same space groups in a different orientation. If in the group $Pmma$, for example, the x and z directions are interchanged the corresponding symbol is $Pcmm$, the glide planes originally parallel to 001 being now normal to the x axis; the further symbols, $Pmcm$, $Pmmb$, $Pbmm$ and $Pmam$ denote the same space group in the remainder of the six orientations possible. In this way, if we pursue the matter to a conclusion, we shall find that the sixty-four different symbols represent only sixteen actual different space groups, as shown in the table below. Highly symmetrical groups such as *Pmmm* and *Pnnn* have the same symbol whatever the orientation of the axes; for others there may be two, three, or six synonymous symbols, and the blanks in each horizontal row are to be filled in by repeating the one, two or three symbols in the row regularly across the columns.

For purposes of indexing and cataloguing, one symbol is selected as corresponding to the ' normal setting ' for each space group, and a set of artificial rules has been established to determine it. The symbols in the first column correspond to this normal setting, but we shall not discuss the rules here since in descriptions of the structures of actual substances we may be called upon to use any one of the possible permutations.

$x\,y\,z$	$z\,y\,x$	$y\,z\,x$	$y\,x\,z$	$z\,x\,y$	$x\,z\,y$
Pmmm	—	—	—	—	—
Pmma	*Pcmm*	*Pmcm*	*Pmmb*	*Pbmm*	*Pmam*
Pmmn	*Pnmm*	*Pmnm*	—	—	—
Pmna	*Pcmn*	*Pncm*	*Pnmb*	*Pbmn*	*Pman*
Pbam	*Pmcb*	*Pcma*	—	—	—
Pban	*Pncb*	*Pcna*	—	—	—
Pbcm	*Pmab*	*Pbma*	*Pcam*	*Pmca*	*Pcmb*
Pbca	*Pcab*	—	—	—	—
Pbcn	*Pnab*	*Pbna*	*Pcan*	*Pnca*	*Pcnb*
Pccm	*Pmaa*	*Pbmb*	—	—	—
Pcca	*Pcaa*	*Pbcb*	*Pccb*	*Pbaa*	*Pbab*
Pccn	*Pnaa*	*Pbnb*	—	—	—
Pnma	*Pcmn*	*Pmcn*	*Pmnb*	*Pbnm*	*Pnam*
Pnnm	*Pmnn*	*Pnmn*	—	—	—
Pnna	*Pcnn*	*Pncn*	*Pnnb*	*Pbnn*	*Pnan*
Pnnn	—	—	—	—	—

When we turn to the consideration of space groups, isomorphous with the point group *mmm*, based on a *C*-face-centred lattice, the restrictions which we have already noticed again operate. There may be *m* planes parallel to 100, which will be interleaved by parallel *b* planes, or there may be *c* planes in this direction interleaved by *n* planes. Parallel to 010, *m* planes involve the presence also of *a* planes, whilst *c* planes will be accompanied by *n* planes. Parallel to 001 we may have *m* planes, but these involve also the operation indicated by

n planes; or a planes, which will also be b planes. The eight possible symbols represent six new space groups—*Cmmm*, *Cmma*, *Cmcm* (= *Ccmm*), *Cmca* (= *Ccma*), *Cccm* and *Ccca*. As in the space groups isomorphous with the point group 222, the similar significance of the x, y and z directions in this point group mean that no new arrangements of symmetry elements can be derived if we base our description on an *A*- or a *B*-face-centred unit; symbols such as *Ammm* or *Bmab* correspond to *C*-face-centred groups with a different axial orientation.

If the lattice be body-centred, m planes parallel to any one of the pinacoids involve also n planes in this direction. Glide planes a, b or c will occur in sets of two kinds of plane, according to the direction in question ; thus b planes parallel to 100 will be accompanied by c planes in this direction also, c planes parallel to 010 by a planes, and so forth. The eight symbols we can derive thus represent only four further space groups—*Immm*, *Imma*, *Imaa*, *Ibca*.

Finally, based on the quadruply-primitive F cell there are two space groups, *Fmmm* and *Fddd*, making a total of twenty-eight space groups isomorphous with the point group *mmm*.

C	I	F
Cmmm	Immm	Fmmm
Cmma	Imma	
Cmcm		
Cmca	Imaa	
Cccm		
Ccca	Ibca	Fddd

It will be remembered that the full symmetry of the class *mmm* includes three diad axes and a centre. All these space groups, therefore, will show families of diads (rotation or screw axes or both) in the x, y and z directions, and also centres of symmetry in addition to the planes of symmetry in terms of which we have symbolised them. To illustrate in terms of a simple example, Fig. 467 shows a part of a pattern based on the space group *Pnma*. The triangular units are repeated to satisfy the operation of the three kinds of plane of symmetry (notice the sign used to denote the presence of a planes parallel to 001), and it will be seen that screw diad axes have arisen in all three axial directions and that there are centres of symmetry along the [010] diad axes.

In drawing patterns of this kind to display the symmetry of space groups we have so far placed the indicating triangles within the unit cell without any special relationship to the symmetry elements—they are in *general positions,* and their repetition displays the number of

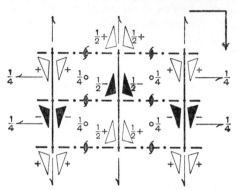

FIG. 467. A portion of a pattern based on the space group *Pnma.*

general equivalent positions of the same kind in each unit of pattern. It must not be supposed, however, that all the atoms in actual crystals occupy such general positions, or that atomic groups are necessarily centred about such positions. Some, or all, of the units may be specially related to the symmetry elements—they may be centred at *special equivalent positions* on axes of symmetry or on planes of sym-

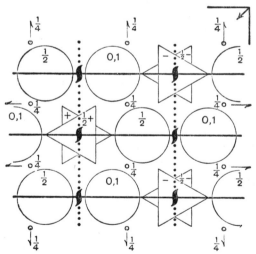

FIG. 468. A *diagrammatic* representation of a portion of the structure of aragonite, with the elements of symmetry indicated.

metry or at centres of symmetry. Fig. 468 represents diagrammatically a portion of the accepted structure of orthorhombic calcium carbonate, aragonite, $CaCO_3$, projected on the $x\,y$ plane. The space group is *Pmcn* ($= Pnma$). The centres of the Ca ions (indicated by circles) lie on the m planes, as also do the centres of the C atoms (in the middle of each equilateral triangle). Of the three oxygens of each CO_3 group, centred at the apices of the equilateral triangles, one lies on an m plane and is thus not repeated by it, but the other two occupy general positions. A unit cell of the structure contains only four $CaCO_3$ equivalents, though there are eight general equivalent points per unit cell in the space group *Pnma*. Notice, however, that whilst the number of equivalent positions is halved, compared with the number of general equivalent positions, for special positions on m planes or on rotation diad axes there is no corresponding reduction for positions on glide planes or on screw diad axes. In aragonite the C atoms are situated on the intersections of the m planes with c planes (Fig. 468), but this does not further reduce the number of such equivalent positions. In the space groups to be studied below we shall encounter three-fold, four-fold and six-fold reductions of the number of equivalent positions for special situations on rotation axes of corresponding degree.

TETRAGONAL SPACE GROUPS

Point group 4.

The vertical tetrad axis of the point group may be paralleled in the space group by axes 4, 4_1, 4_2 or 4_3, whilst the space lattice in question may be tetragonal *P* or tetragonal *I*. The translation $\frac{1}{2}\frac{1}{2}\frac{1}{2}$ involved in the doubly-primitive *I* cell, however, gives rise to 4_2 axes parallel to the rotation tetrad axes in the space group *I*4 (Fig. 469), and to 4_3 axes parallel to the 4_1 axes in the group *I*4_1. (Notice also in Fig. 469 that diad axes arise parallel to the tetrad axes.) Hence there are only six distinct space groups isomorphous with the point group 4:

P	*I*
*P*4	*I*4
P4_1	*I*4_1
P4_2	
P4_3	

The distribution of the various kinds of tetrad axes in these space groups may help to explain the observations we have made concerning the phenomenon of optical activity exhibited by some substances crystallising in such an enantiomorphous class. If the structure is based on a space group containing screw axes of one kind only, as in the group $P4_1$ or the enantiomorphous group $P4_3$, then it would seem highly probable that the appropriate physical characteristics would be present which result in the substance being optically active. A structure based on such a space group as $P4$ or $P4_2$, however, seems less likely necessarily to possess these characteristics and so may build up a crystal which, while evidently correctly assigned to the enantiomorphous class 4, does not show optical activity (cf. p. 154).

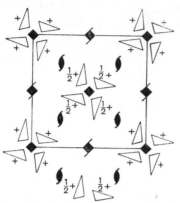

FIG. 469. A portion of a pattern based on the space group $I4$.

Point group $\bar{4}$.

Since there is only the possibility $\bar{4}$ for the vertical axes, there are only two space groups, $P\bar{4}$ and $I\bar{4}$, isomorphous with this point group.

Point group $4/m$.

With the introduction of horizontal planes of symmetry, vertical axes with a definite sense of screw (4_1 or 4_3) would be converted by reflection to their enantiomorphs (4_3 or 4_1 respectively). Hence in all space groups isomorphous with the point group $4/m$, either the vertical tetrad axes will be axes 4 or 4_2 (without sense of screw), or axes 4_1 and 4_3 will occur in parallel sets. If the horizontal planes are m planes, only axes 4 and 4_2 are possible, giving the groups $P4/m$ and $P4_2/m$ based on a P lattice. The same is true if the planes are n planes, for the translation by an n plane brings one vertical axis into coincidence with another of the same kind (see the arrangement of axes in Fig. 469), so that there arise two further groups $P4/n$ and $P4_2/n$. In the group $I4/m$ both m planes and n planes occur interleaved.

A final possibility may best be envisaged by remembering that the I cell may be described as an all-face-centred F cell by an alternative choice of x and y axial directions at 45° to the original. In the two sets of tetragonal space groups described above, this alternative choice merely gives rise to synonymous symbols—$F4$ for $I4$, $F\bar{4}$ for $I\bar{4}$, and so on. Planes of symmetry parallel to the centred 001 face of such a cell, however, may be d planes (p. 245), and so there arises a new arrangement $F4_1/d$ (conventionally described as $I4_1/a$ to correspond to the original setting), in which axes 4_1 and 4_3 occur in parallel sets. There are six space groups in all isomorphous with the point group $4/m$.

P	I
$P4/m$	$I4/m$
$P4_2/m$	
$P4/n$	
$P4_2/n$	$I4_1/a$

Point group 4*mm*.

The first letter m of the point group symbol denotes the principal planes of symmetry, parallel to the faces of the {100} tetragonal prism. In space groups based on a primitive space lattice the planes parallel to 100 may be m, b, c or n planes. The second letter m of the symbol refers to the diagonal planes of symmetry at 45° to the crystallographic x, y directions. If we bear in mind that the tetragonal P space lattice can also be described in terms of a C face-centred cell with its x, y axes taken in the [110] directions of the P cell, it will be clear that there are fewer possibilities for the second set of planes in the isomorphous space groups. Diagonal m planes will involve also the presence of interleaved glide planes with a glide component normal to the tetrad axes, whilst c planes in the diagonal positions will be interleaved by parallel n planes. Hence eight distinct groups based on a P lattice exist.

In groups based on the tetragonal I space lattice, m planes in the principal positions will be interleaved by parallel n planes, and c planes in this position by b planes. Planes in the diagonal positions are parallel to the centred faces of an F cell, and thus may be m planes or d planes. Four further space groups thus result, giving a total of twelve groups isomorphous with the point group 4*mm*.

P	I
P4*mm*	I4*mm*
P4*bm*	
P4*cm*	I4*cm*
P4*nm*	
P4*mc*	I4*md*
P4*bc*	
P4*cc*	I4*cd*
P4*nc*	

The above table gives the original symbols for these space groups. Each group is sufficiently defined by the letters indicating the nature of the two kinds of planes of symmetry. If, however, we develop a pattern based, for example, on a tetragonal P lattice and the planes cm we find that the only kind of tetrad axes which arise are axes 4_2; on the 'rule of priority' adopted in other point groups the symbol should read $P4_2cm$. The symbols in the following table show the highest kind of tetrad axes present in each group, and are to be preferred.

Axes 4	Axes 4_2	Axes 4_1 and 4_3
P4mm	P4$_2$cm	I4$_1$md
P4bm	P4$_2$nm	I4$_1$cd
P4cc	P4$_2$mc	
P4nc	P4$_2$bc	
I4mm		
I4cm		

Point group $\bar{4}2m$.

The order of the symbols in $\bar{4}2m$ indicates that, having set the inversion tetrad axis vertical, the crystal is orientated so that the horizontal diad axes can be chosen as the x and y directions, the planes of symmetry thus being set diagonally. In space groups based on a P lattice the diad axes may be rotation axes 2 or screw axes 2_1, whilst, as in the discussion above, only m or c arise as possibilities for the symbol of the associated planes. Four new space groups can thus be derived. Instead of setting the planes diagonally in the P cell, however (Fig. 470), they may be set parallel to the {100} faces of this cell (Fig. 471),

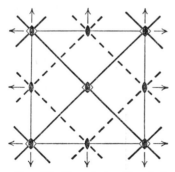

FIG. 470. The relationship of the planes of symmetry to the outline of the P cell in the space group $P\bar{4}2m$.

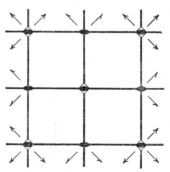

FIG. 471. The relationship of the planes of symmetry to the outline of the P cell in the space group $P\bar{4}m2$.

and hence diagonally in the alternative C cell. We might describe such space groups in terms of the C cell, but it is simpler to retain the primitive cell and to indicate the new setting, with the x and y crystallographic axes normal to the planes of symmetry, by writing the symbol $\bar{4}m2$ instead of $\bar{4}2m$. Planes in the space groups in this attitude may be m, b, c or n planes; rotation diad axes in the diagonal position, parallel to the [100] and [010] directions of a C cell, will be accompanied by sets of parallel 2_1 axes so that only rotation diad axes 2 will figure in the symbols. We thus derive four further new space groups based on a primitive cell (see the table below).

In space groups based on an I cell planes set in the attitude $\bar{4}2m$ are parallel to the {100} faces of an F cell, and so may be m planes (accompanied by c planes) or d planes. In the setting $\bar{4}m2$ they may be m planes (paralleled by n planes) or c planes (paralleled by b planes). Only axes 2 will appear in the symbols. We thus have derived twelve distinct space groups isomorphous with this point group.

P	I
$P\bar{4}2m$	$I\bar{4}2m$
$P\bar{4}2_1m$	
$P\bar{4}2c$	$I\bar{4}2d$
$P\bar{4}2_1c$	
$P\bar{4}m2$	$I\bar{4}m2$
$P\bar{4}b2$	
$P\bar{4}c2$	$I\bar{4}c2$
$P\bar{4}n2$	

Point group 42.

The derivation of space groups isomorphous with this point group involves little difficulty. In groups based on a P cell the vertical axes may be any one of four possibilities, 4, 4_1, 4_2 or 4_3, whilst the diad axes parallel to the x and y directions may in each case be either rotation diads 2 or screw diads 2_1, so that there are eight different groups. Notice that some of these, such as $P4_12_1$ and $P4_32_1$, constitute enantiomorphous pairs. In all of them, of course, further diad axes arise in the diagonal directions, and since these are the x and y directions of a C cell both kinds of axes 2 and 2_1 will be present in all groups. With an I cell vertical axes 4 will be accompanied by a parallel set of 4_2 axes, and vertical axes 4_1 by parallel 4_3 axes, precisely as in the space groups isomorphous with the point group 4; the horizontal axes will consist of sets of rotation diad axes parallel to sets of screw diad axes in both [100] and [110] directions, so that only two further space groups arise.

P	I
$P42$	$I42$
$P42_1$	
$P4_12$	$I4_12$
$P4_12_1$	
$P4_22$	
$P4_22_1$	
$P4_32$	
$P4_32_1$	

Point group 4/mmm.

Just as the crystal class 4/*mmm* may be derived from the class **4mm** by the introduction of a plane of symmetry normal to the tetrad axis, so we can proceed most easily here by introducing sets of horizontal symmetry planes into the twelve space groups isomorphous with the point group **4mm** (p. 255). We know already (p. 254) that in the P groups these may be *m* planes or *n* planes, whilst in the I groups planes of type *m* or *a* are involved. There are twenty space groups isomorphous with the point group 4/*mmm*. In all these groups there are, of course, centres of symmetry, additional axes $\bar{4}$, 4_2, 4_1, 4_3, 2 or 2_1 parallel to the vertical tetrad axes, and horizontal diad axes in both kinds of direction. In the original symbols, as in **4mm** groups, the vertical tetrad axes are indicated by the figure 4 whatever the particular kind of tetrads actually present.

P		I
$P4/mmm$	$P4/nmm$	$I4/mmm$
$P4/mbm$	$P4/nbm$	
$P4/mcm$	$P4/ncm$	$I4/mcm$
$P4/mnm$	$P4/nnm$	
$P4/mmc$	$P4/nmc$	$I4/amd$
$P4/mbc$	$P4/nbc$	
$P4/mcc$	$P4/ncc$	$I4/acd$
$P4/mnc$	$P4/nnc$	

The following table (compare p. 256) shows, in accord with the ' rule of priority ', the nature of the highest kind of tetrad axes present, and these symbols are to be preferred (see p. 273).

Axes 4	Axes 4_2	Axes 4_1 and 4_3
$P4/mmm$	$P4_2/mcm$	$I4_1/amd$
$P4/nmm$	$P4_2/ncm$	$I4_1/acd$
$P4/mbm$	$P4_2/mnm$	
$P4/nbm$	$P4_2/nnm$	
$P4/mcc$	$P4_2/mmc$	
$P4/ncc$	$P4_2/nmc$	
$P4/mnc$	$P4_2/mbc$	
$P4/nnc$	$P4_2/nbc$	
$I4/mmm$		
$I4/mcm$		

TRIGONAL SPACE GROUPS

We have already noticed on several occasions the close relationship which exists between the trigonal and hexagonal systems, and this is again brought out when we begin to study the symmetry of the underlying structure. Some trigonal crystals are built on a rhombohedral space lattice, for which, as we have seen, we can always select a primitive unit cell. We might denote this the *rhombohedral P cell*, but because of other considerations involved the lattice has been given the special symbol *R*. Other trigonal crystals, crystallising perhaps in the same class, are found to be built on a hexagonal space lattice precisely the same as the lattice appropriate to truly hexagonal substances. We could describe this lattice also in terms of a rhombohedral unit, but it would necessarily be triply primitive (Figs. 472, 473), and it is simpler

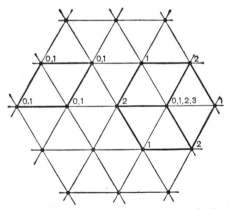

FIG. 472. Basal projection of part of a hexagonal space lattice, showing the hexagonal *C* cell and the triply primitive rhombohedral cell.

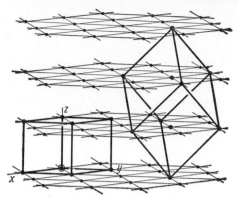

FIG. 473. Clinographic view of the unit cells shown in Fig. 472

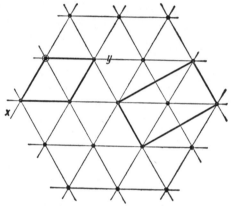

FIG. 474 Basal projection showing the relationship of the hexagonal C cell to an orthogonal unit cell.

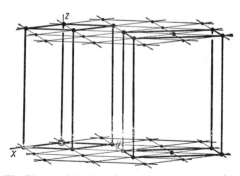

FIG. 475. Clinographic view of the unit cells shown in Fig. 474

to use the hexagonal unit which we have already figured (p. 226). This unit is a right prism based on a rhombus with edges parallel to the x and y crystallographic directions. It is itself primitive, but if orthogonal axes were chosen (Fig. 474) the corresponding unit cell would be C face-centred and so it was originally given the symbol C. We shall continue to use this symbol for the present, but it has been agreed to replace it by P (see p. 273). The relationship of the C cell to an orthogonal cell is shown in clinographic view in Fig. 475.

Point group 3.

In space groups based on the C space lattice the triad axes may be all rotation triad axes 3, screw axes 3_1 or screw axes 3_2. The R lattice, however, itself involves translations of $\frac{1}{3}$ and $\frac{2}{3}$ parallel to the vertical

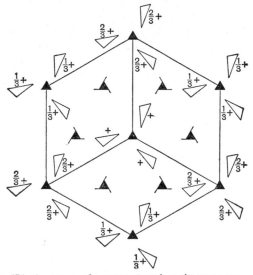

FIG. 476. A portion of a pattern based on the space group $R3$.

axis, so that the space group $R3$ has sets of 3_1 axes and 3_2 axes parallel to the rotation triad axes (Fig. 476). There are thus four distinct space groups isomorphous with this point group.

C	R
$C3$	$R3$
$C3_1$	
$C3_2$	

Point group $\bar{3}$.

The only possibilities here are $C\bar{3}$ and $R\bar{3}$. Both space groups, of course, include centres of symmetry; in $C\bar{3}$ only rotation triad axes are present ($\bar{3}$ being equivalent to a rotation triad axis together with a centre, p. 104), whilst in $R\bar{3}$ there are all three kinds of axes 3, 3_1 and 3_2.

Point group 3*m*.

In morphological descriptions of crystals belonging to class 3*m* we have been accustomed so to set the crystal that the *x* and *y* directions

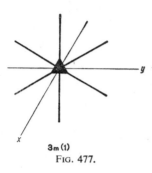

3m (1)

Fig. 477.

are normal to planes of symmetry (Fig. 477). In the isomorphous space groups the vertical symmetry planes may be *m* planes or *c* planes, giving space groups $C3m$ and $C3c$ based on the C lattice; remembering the translations involved in an orthogonal C cell, we shall realise that there are sets of glide planes parallel to the *m* planes of the group $C3m$, with a horizontal glide component, and sets of *n* planes parallel to the *c* glide planes of the group $C3c$. Instead of setting the symmetry planes in the space group normal to the *x* and *y* axes of the hexagonal C cell, however, we may build up further groups by introducing planes parallel to the vertical faces of the cell. These directions are normal to the edges of a possible triply primitive unit cell

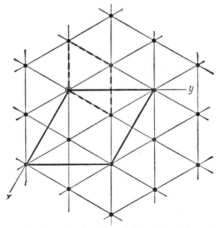

Fig. 478. Basal projection showing the relationship of the triply primitive *H* cell to the hexagonal *C* cell.

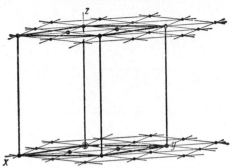

FIG. 479. A clinographic view of the *H* cell.

(Figs. 478, 479), to which the symbol *H* has been given, and we might denote the corresponding space groups *H3m* amd *H3c*. More simply, we can adopt a device (similar to the one which we used in the point group $\bar{4}2m$) to indicate the choice of *x* and *y* axes in the planes of symmetry instead of normal to these planes, and write 31*m* (Fig. 480) instead of 3*m*(1), the figure 1 referring to identity axes in the corresponding directions; the symbol *H* is then unnecessary. The two new space groups which we have derived are then written *C31m* and *C31c*.

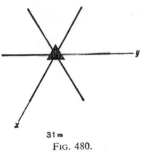

31 m

FIG. 480.

Fig. 481 shows in plan a part of the space group *C3m* with the base of the *C* cell outlined, whilst Fig. 482 shows the relationship of the

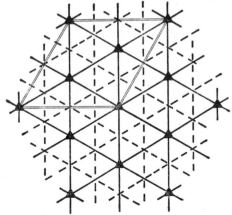

FIG. 481. A portion of the space group *C3m*. The base of the *C* cell is outlined by double lines.

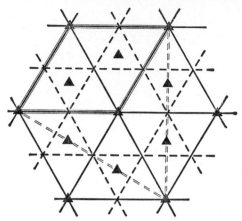

FIG. 482. A portion of the space group $C31m$. The base of the C cell is outlined by double lines, whilst broken double lines show the direction of the x and y axes of the H cell.

symmetry planes to the C cell and to the axes of the H cell (dashed lines) in the group $C31m$.

Two further groups arise, based on the R lattice, for which we may write the symbols $R3m$ and $R3c$. A hexagonal cell is sometimes used in

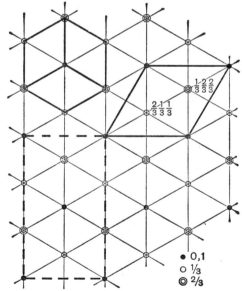

FIG. 483. Basal projection of part of a rhombohedral space lattice showing the R cell, hexagonal cell and (broken lines) the sextuply primitive orthogonal cell.

describing these groups, the points $\frac{1}{3}\frac{2}{3}\frac{2}{3}$ and $\frac{2}{3}\frac{1}{3}\frac{1}{3}$ being equivalent to the point 0 0 0, so that the cell is triply primitive (Figs. 483, 484). Fig.

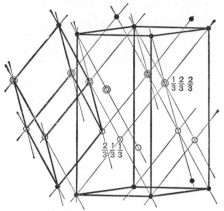

FIG. 484. Clinographic views of the R cell and hexagonal cell of Fig. 483.

483 shows also in dashed outline the base of the smallest orthogonal cell, a sextuply primitive one, which could be used in place of the R cell. Of these we shall only need at present to use the R cell, together with the C cell described above.

We may now tabulate the six space groups isomorphous with the point group $3m$.

C	R
$C3m$	$R3m$
$C3c$	$R3c$
$C31m$	
$C31c$	

Point group $\bar{3}m$.

The standard orientation in this crystal class involves the choice of two of the diads as x and y directions, the planes of symmetry thus being normal to the crystallographic axes as in class $3m$ (Fig. 485). On the C lattice we may build up space groups $C\bar{3}m$ and $C\bar{3}c$. The group $C\bar{3}m$ includes glide planes with a horizontal component of glide interleaved between the m planes, diad axes 2 and 2_1 (remember the translations involved in a C cell) normal to these planes, and centres of symmetry. In $C\bar{3}c$ the c planes alternate with n planes, and centres and both kinds of diad axes are again present.

As with space groups isomorphous with the point group $3m$, we may construct further groups by placing the planes of symmetry parallel to the faces of the C cell instead of normal to them. To indicate this change of orientation of the point group in relation to the x and y

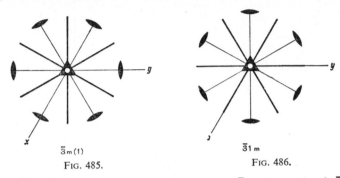

$\bar{3}m(1)$

FIG. 485.

$\bar{3}1m$

FIG. 486.

axes (Fig. 486), we must write the symbol $\bar{3}1m$ instead of $\bar{3}m(1)$. (Notice particularly that the symbol is *not* written $\bar{3}2m$; a figure 2 in the second place in the symbol would denote the presence of diad axes in the *secondary* positions (parallel to the x and y directions), as it does in $\bar{4}2m$. In the new orientation here in question the x and y directions are identity axes, and we must symbolise this by the figure 1, as in $\bar{3}1m$). Two further space groups $C\bar{3}1m$ and $C\bar{3}1c$ arise, and finally on the R lattice we may build two more, $R\bar{3}m$ and $R\bar{3}c$.

C	R
$C\bar{3}m$	$R\bar{3}m$
$C\bar{3}c$	$R\bar{3}c$
$C\bar{3}1m$	
$C\bar{3}1c$	

Point group 32.

Similar considerations arise in building up space groups isomorphous with this point group. In morphological studies the x and y directions are chosen parallel to diad axes (Fig. 487) as denoted by the symbol 32. In the isomorphous space groups the triad axes may be axes 3, 3_1 or 3_2, axes of one kind only being present in each of the space groups $C32$, $C3_12$ and $C3_22$, whilst all three kinds of triad are present in the space group $R32$. In the orientation 312 the horizontal diad axes are normal to the x and y directions (Fig. 488), and we have the further space groups $C312$, $C3_112$ and $C3_212$. In all seven space

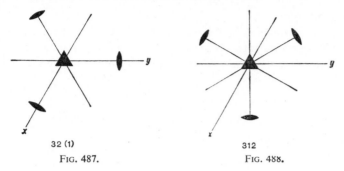

32 (1)

FIG. 487.

312

FIG. 488.

groups the rotation diad axes alternate with screw diad axes, but there are, of course, no centres of symmetry. Space groups such as $C3_12$ and $C3_22$ constitute enantiomorphous pairs.

C	R
$C32$	$R32$
$C3_12$	
$C3_22$	
$C3\bar{1}2$	
$C3_1\bar{1}2$	
$C3_2\bar{1}2$	

HEXAGONAL SPACE GROUPS

All hexagonal crystals are built upon a truly hexagonal lattice, and the C cell which we have used above affords the most convenient basis for description of the isomorphous space groups.

Point group 6.

The vertical axes in the isomorphous space groups may be rotation axes 6 or any one of the five kinds of screw hexad axes which we have already described. In each space group only one kind of hexad axis occurs, but the six-fold axes are paralleled by sets of triad axes 3, 3_1 or 3_2 and sets of diad axes 2 or 2_1. It need scarcely be pointed out that no planes of symmetry and no centres of symmetry are present. The six space groups are symbolised $C6$, $C6_1$, $C6_2$, $C6_3$, $C6_4$ and $C6_5$.

Point group $\bar{6}$.

Only one space group, $C\bar{6}$, can be built up here; it includes, clearly, a set of reflection planes normal to the inversion hexad axes, since the crystal class $\bar{6}$ is equivalent to $3/m$.

Point group 6/m.

Two space groups arise, $C6/m$ and $C6_3/m$. In both there are rotation triad axes 3, either rotation diad axes 2 or screw diad axes 2_1 parallel to the vertical hexad axes, and centres of symmetry along the axes of even degree.

Point group 6mm.

Each of the two sets of vertical planes may be reflection planes m or glide planes c, giving four space groups $C6mm$, $C6mc$, $C6cm$ and $C6cc$. Glide planes with a horizontal glide component will alternate with m planes in either position, whilst c planes will be interleaved by n planes. Since in two of these groups the only hexad axes present are axes 6_3 it is preferable to use the symbols $C6_3mc$ and $C6_3cm$ respectively for these groups (see p. 273).

Point group $\bar{6}m2$.

Setting the planes of symmetry in this point group in the conventional attitude normal to the x and y crystallographic axes (Fig. 489), the diad axes are in the *tertiary* positions(normal to the crystallographic axes), and not in the secondary positions as they were in the group $\bar{3}m$. This fact is indicated by the figure 2 appearing in the third place in the symbol; on the C space lattice we build up the space groups $C\bar{6}m2$

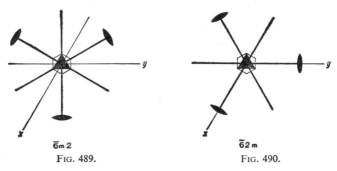

$\bar{6}m2$
FIG. 489.

$\bar{6}2m$
FIG. 490.

and $C\bar{6}c2$. If the planes of symmetry are set parallel to the vertical faces of the C cell, the diad axes coincide with the x, y directions (Fig. 490); they occupy the secondary positions, and we write the symbol $\bar{6}2m$. The further space groups resulting are $C\bar{6}2m$ and $C\bar{6}2c$. As before, glide planes with a horizontal glide component alternate with m planes, and n planes alternate with c glide planes; there are sets of both rotation diad and screw diad axes in all four groups, and reflection planes normal to the inversion hexad axes.

Point group 62.

Six possibilities clearly arise here, denoted $C62$, $C6_12$, $C6_22$, $C6_32$, $C6_42$ and $C6_52$. There are sets of screw diad axes parallel to the rotation diads in all these groups. $C6_12$ and $C6_52$ are enantiomorphous pairs, as also are $C6_22$ and $C6_42$.

Point group 6/*mmm*.

Only reflection planes *m* arise as possibilities normal to the hexad axis, and we can derive the four new space groups from those groups isomorphous with the point group 6*mm* by the addition of such planes. They are the groups $C6/mmm$, $C6/mmc$, $C6/mcm$ and $C6/mcc$ (the second and third preferably symbolised $C6_3/mmc$ and $C6_3/mcm$ respectively).

CUBIC SPACE GROUPS

Study of the cubic space groups in detail involves rather special difficulties, since the obliquity of axes and planes of symmetry to the crystallographic axial planes makes their portrayal on paper decidedly awkward. We shall not attempt a rigid discussion here, but rather content ourselves with a brief outline of the number of different space groups which arise.

Point group 23.

The space groups isomorphous with this point group show a close analogy with those isomorphous with the group 222 (p. 249), the orthorhombic cell being specialised so that all three axial directions are equivalent. We may have a group $P23$, with rotation diad axes only, and a group $P2_13$ with screw diad axes only. No arrangements with cubic symmetry are possible based on a *C* face-centred unit. The groups $I23$ and $F23$ include sets of screw diad axes parallel to the rotation diad axes. One further group arises, however, by specialisation of the orthorhombic group $I2_12_12_1$. Like this orthorhombic group it contains sets of non-intersecting screw diad axes parallel to sets of non-intersecting rotation diad axes, and we must again forego our 'rule of priority', and denote the group by the symbol $I2_13$ in spite of the presence in it of rotation diad axes. There are thus five space groups isomorphous with this point group.

P	I	F
$P23$	$I23$	$F23$
$P2_13$	$I2_13$	

Point group m3.

In groups based on the cubic *P* lattice, the planes of symmetry may be *m* planes, glide planes with an axial glide component, or *n* glide planes. The groups *Pm3*, *Pa3* and *Pn3* are specialisations of the orthorhombic groups *Pmmm*, *Pbca* and *Pnnn* (p. 250). By specialising the orthorhombic groups *Immm* and *Ibca* we derive the cubic groups *Im3* and *Ia3*, and from *Fmmm* and *Fddd* there arise groups *Fm3* and *Fd3*. It should be unnecessary to particularise the further sets of planes of symmetry which arise in the groups based on *I* and *F* cells, or the diad axes and centres of symmetry present in all the groups.

P	*I*	*F*
Pm3	*Im3*	*Fm3*
Pa3	*Ia3*	
Pn3		*Fd3*

Point group 4̄3m.

We may derive the space groups here by specialisation of those groups isomorphous with the point group 4̄2m in which *x* and *y* are directions of diad axes, remembering that the groups *I4̄m2* and *I4̄c2* (p. 257) may be written also *F4̄2m* and *F4̄2c* respectively. Since the planes of symmetry in these space groups are in diagonal positions, there will be an alternation of two kinds of plane even in those groups based on a primitive lattice.

P	*I*	*F*
P4̄3m	*I4̄3m*	*F4̄3m*
P4̄3c	*I4̄3d*	*F4̄3c*

(The group *P4̄3c* is usually written *P4̄3n*. The two types of glide plane are interleaved, but the letter *c* strictly applies only to the two sets of glide planes gliding parallel to [001], other members of this same family, gliding parallel to [100] and to [010], being *a* planes and *b* planes respectively. The interleaved diagonal planes can be denoted by *n* whatever their orientation.)

Point group 43.

The space groups isomorphous with this point group are specialisations of tetragonal groups isomorphous with the point group 42. In all of them there are sets of rotation diad axes alternating with screw diad axes in the [110] directions.

P	I	F
$P43$	$I43$	$F43$
$P4_13$	$I4_13$	$F4_13$
$P4_23$		
$P4_33$		

Point group m3m.

These groups will arise by specialisation of tetragonal space groups isomorphous with the point group $4/mmm$, of which we need consider only those (p. 258) in which the planes parallel to 100 and to 001 are of the same kind. From groups $P4/mmm$, $P4/mmc$, $P4/nnm$ and $P4/nnc$ respectively we derive cubic groups $Pm3m$, $Pm3c$, $Pn3m$ and $Pn3c$. From $I4/mmm$ ($=F4/mmm$) there arise $Im3m$ and $Fm3m$; from $I4/mcm$, ($=F4/mmc$), is derived $Fm3c$; from $I4/amd$ ($=F4/ddm$) is derived $Fd3m$; from $I4/acd$ ($=F4/ddc$) cubic groups $Ia3d$ and $Fd3c$.

P	I	F
$Pm3m$	$Im3m$	$Fm3m$
$Pm3c$		$Fm3c$
$Pn3m$		$Fd3m$
$Pn3c$	$Ia3d$	$Fd3c$

(The groups $Pm3c$ and $Pn3c$ are conventionally written $Pm3n$ and $Pn3n$ respectively; compare the choice of the symbol $P\bar{4}3n$ in place of $P\bar{4}3c$, p. 270.)

We have now outlined briefly the derivation of the 230 different types of structural pattern on which the structures of actual crystals may be based. The notation which we have used is based upon one evolved by Prof. Ch. Mauguin of the University of Paris and Dr. C. Hermann of the University of Stuttgart; it was used in an atlas of the space groups published in *International Tables for the Determination of Crystal Structures* (1935). As an accepted standard these have now been replaced by *International Tables for X-ray Crystallography*, vol. I Symmetry Groups, 1952, in which a few minor changes in space group symbols have been made. Some of these have been suggested in the

THE DISTRIBUTION OF THE 230 SPACE GROUPS BETWEEN THE 32 CLASSES

TRICLINIC SYSTEM		TRIGONAL SYSTEM	
Class 1 $1\rbrace 2$		Class 3 4	
„ $\bar{1}$ 1		„ $\bar{3}$ 2	
		„ $3m$ $6\rbrace 25$	
MONOCLINIC SYSTEM		„ $\bar{3}m$ 6	
		„ 32 7	
Class 2 3			
„ m $4\rbrace 13$		HEXAGONAL SYSTEM	
„ $2/m$ 6			
		Class 6 6	
ORTHORHOMBIC SYSTEM		„ $\bar{6}$ 1	
		„ $6/m$ 2	
Class $mm2$ 22		„ $6mm$ $4\rbrace 27$	
„ 222 $9\rbrace 59$		„ $\bar{6}m2$ 4	
„ mmm 28		„ 622 6	
		„ $6/mmm$ 4	
TETRAGONAL SYSTEM			
		CUBIC SYSTEM	
Class 4 6			
„ $\bar{4}$ 2		Class 23 5	
„ $4/m$ 6		„ $m3$ 7	
„ $4mm$ $12\rbrace 68$		„ $\bar{4}3m$ $6\rbrace 36$	
„ $\bar{4}2m$ 12		„ 432 8	
„ 422 10		„ $m3m$ 10	
„ $4/mmm$ 20			

preceding pages, and they are all such that they will be readily understood by a student of this chapter. In the *monoclinic system*, in which we have always set the unique axis in the y position, an alternative setting is recognised in which it is set in the z position; thus the space group Cc has an alternative description Bb, the group $P2/c$ an alternative $P2/b$ and so forth. In the *orthorhombic system*, the point symbol group mm is now to be written in full as $mm2$, the unique axis being always set in the z position; the symbols of the isomorphous space groups are likewise written in full, so that we have groups $Pmm2$, $Cmc2_1$ etc. In the *tetragonal system*, the full symbol 422 is used for the trapezohedral class, and a number of symbols have been modified in two other classes to indicate the nature of the tetrad axes in the manner suggested in the tables on p. 256 and p. 259 above. In the *trigonal* and *hexagonal* systems, the *H* setting (p. 263) is no longer used and the symbol *C* is to be replaced by *P* as we have already noted (p. 261); parallel changes are also made with those in the tetragonal system, the symbol 62 being written in full as 622 and the nature of the hexad axes being indicated in four space groups as we have proposed on pp. 268–9. Finally, in the cubic system, a parallel modification gives 432 as the accepted point group symbol for the pentagonal icositetrahedral class, with corresponding expansion of the space group symbols to *P*432 etc. A note on the older notation employed by Schoenflies and still extensively used will be found in the appendix (p. 312). The table on p. 272 shows the distribution of the space groups between the 32 crystal classes, for which we now use the new standard symbols.

CHAPTER XII

DIFFRACTION OF X-RAYS BY CRYSTALS

With the publication of the work of Schoenflies, Fedorov and Barlow the geometrical theory of crystal structure was substantially complete at the end of the last century. This work, however, afforded no indication whether stable crystals could in fact be built up on all these patterns, nor was it possible to determine the actual space group appropriate to a given crystalline substance. Attempts were made, it is true, to deduce the probable space groups in certain cases: quartz, crystallising in enantiomorphous modifications in class 32 and displaying optical activity, was naturally assigned to the groups $C3_12$ and $C3_22$ or to the groups $C3_112$ and $C3_212$, and Sohncke argued in favour of the former alternative; calcite was referred to the space group $R\bar{3}c$. An unambiguous determination of the space group on which the structural pattern of a given crystal is based first became possible with the development of crystal structure analysis by means of X-rays which followed rapidly upon Laue's discovery (p. 155).

At the time of this experiment in 1912 it was still uncertain whether X-rays were corpuscular or undulatory in character. In the University of Munich a group of physicists were working on problems of the transmission of light through crystals, and in the course of discussion the suggestion was made that if X-rays also were undulatory then the regular spacing of units in a crystal should be of the appropriate order of magnitude to effect diffraction of an X-ray beam. The first experiments were actually carried out by Friedrich and Knipping. A beam of X-rays was passed through a crystal of copper sulphate and received on a photographic plate: after a few trial experiments, a plate showed on development a central black area due to the direct beam surrounded by a number of black spots due to diffracted beams. To the physicist the great importance of this result lay in the proof which further experiments soon afforded of the wave-character of X-rays and the first approximate estimates of wavelength to which they led. The crystallographer, on the other hand, was presented with a proof of the regularity of the crystal structure and with the means of analysing this structure directly.

In the Laue method a beam of ' white ' X-rays is used, covering a range of wavelengths.   This method is still useful for the study of crystal symmetry after the manner briefly outlined on pp. 155–6, but in later work there have been developed a number of other practical procedures, found to be more generally useful in detailed structural analysis, in which the beam is rendered monochromatic.   By using an appropriate metal in the target of an X-ray tube excited at the correct voltage, and transmitting through a filter, the resultant beam is practically restricted to radiation of a single wavelength, the value depending on the particular metal employed.   Thus for cobalt the filtered beam has $\lambda = 1\cdot79$ Ångström Units, for molybdenum $\lambda = 0\cdot71$ A.

When a crystal is placed in the path of such a beam its atoms act, owing to the forced vibrations induced in the electrons, as secondary sources emitting X-rays.   The frequency and wavelength of these emitted rays are identical with those of the incident beam.   In general, the crystal may be said to ' scatter ' the X-rays, but by analogy with optical diffraction we may expect that in certain directions the individual scattered wavelets may recombine in phase to produce a strong reinforced but deviated beam (Fig. 491).   For a single row of atoms, as in this figure, the necessary condition for reinforcement is that $\angle IBA = \angle RBC$—the primary and secondary beams make equal angles with the

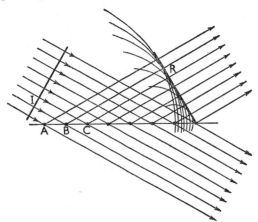

Fig. 491. Reinforced diffraction from a point row.

line row of atoms, which thus appears to be ' reflecting ' the beam. This reflection analogy was first pointed out by W. L. Bragg, and it is now quite customary for X-ray crystallographers to talk of the reflection of X-rays by crystals.   The beam of X-rays, however, penetrates the crystal in such a manner that atoms in many successive rows (successive planes in the three-dimensional crystal) also become secondary sources. A further condition must be fulfilled to produce a final comprehensive reinforced reflection. From Fig. 492 the path difference for the two rays shown

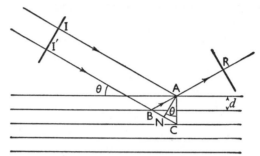

FIG. 492. Reinforced diffraction from a series of planes.

$$I'BR - IAR = I'B + BA - IA$$
$$= I'N + NC - IA$$
$$= NC$$
$$= 2d \sin \theta$$

where $d =$ the *spacing* of the structure planes and $\theta =$ the ' glancing angle ' (the complement of the optical angle of incidence and reflection). For reinforcement this path difference must equal $n\lambda$, where n is an integer and λ the wavelength of the incident beam. We thus reach the conclusion that a strong ' reflected ' beam is produced when the Bragg equation

$$n\lambda = 2d \sin \theta$$

is satisfied. The value of n determines, by analogy with optical diffraction, the *order* of the reflection; an upper limit is set, in any given case, by the fact that $\sin \theta = \dfrac{n\lambda}{2d}$ cannot have a value greater than unity.

One of the most frequently used techniques is that of the *rotation photograph*. The crystal is set up in the path of the X-ray beam so that it can be continuously rotated about a prominent zone axis (in the customary arrangement of the apparatus the X-ray beam is horizontal and the axis of rotation of the crystal vertical). The diffracted rays fall on a cylindrical film placed coaxially with the axis of rotation. Let us suppose that an orthorhombic crystal is rotated about its crystallographic z axis. As rotation proceeds, various families of $hk0$ planes are brought in turn into angular positions such that the equation $n\lambda = 2d \sin \theta$ is momentarily satisfied by the particular values of the spacing d and the glancing angle θ for the family of planes in question. An intense diffracted beam is thus produced at a deviation of 2θ on one or the other side of the direct beam in a horizontal plane through the beam (Fig. 493). The diffraction is recorded on the film as a series

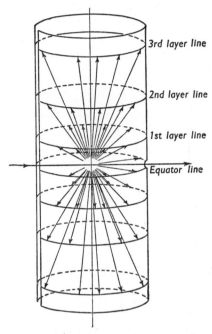

FIG. 493. Diagram of apparatus for rotation photograph.

of darkened spots on the horizontal *equator line*; each momentary appearance of a particular diffracted beam, as the crystal passes through the appropriate position, will produce only a very slight darkening of the photographic emulsion, but the apparatus is left to run for a period sufficiently long to produce a photograph of the required density.

Planes $h\ k\ l$ produce during rotation similar diffracted beams directed along the surfaces of a series of cones coaxial with the axis of rotation. These cones intersect the cylindrical film in a series of horizontal circles (Fig. 493), so that when the film is unrolled and developed a further series of spots mark successive *layer lines* parallel to, and above and below, the equator line (Fig. 494). Spots on the first layer line must

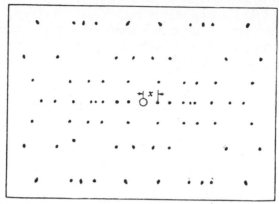

FIG. 494. Diagrammatic reproduction of a rotation photograph.

correspond to diffracted beams from successive identically-situated atoms in the structure experiencing a path-difference of one whole wavelength. Thus, in Fig. 495 if c_0 is the *identity period* along the axis

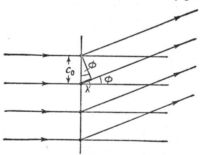

FIG. 495. Elevation of diffracted rays to produce the first layer line.

of rotation (i.e. the length of side of the unit cell in this direction), the angle of elevation ϕ of the diffracted beam is given by

$$\sin \phi = \frac{\lambda}{c_0}.$$

If we measure the height h_1 of the first layer line above the equator line on the photograph, and the radius of the camera is r, then

$$\tan \phi = \frac{h_1}{r}$$

(Fig. 496), whence by elimination of ϕ we can calculate c_0. The measurements can be made more accurately using a higher layer line, so that in general for the nth layer line

$$c_0 = \frac{n\lambda}{\sin\left(\tan^{-1}\dfrac{h_n}{r}\right)}$$

(the distance actually measured in practice is the double distance $2h_n$

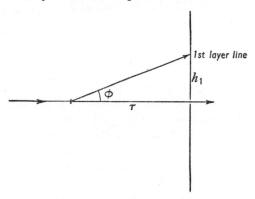

1st layer line

h_1

ϕ

r

FIG. 496. Diagram showing the relationship between the height of a layer line and the radius of the camera.

across the equator line). Thus we are able for the first time to make an absolute measurement of the dimensions of the unit cell of a crystal (for the orthorhombic crystal under consideration, two further rotation photographs about the x and y crystallographic axes respectively would yield the remaining dimensions a_0 and b_0).

Unit cell dimensions thus absolutely determined invite comparison with axial *ratios* determined morphologically after selection of a parametral plane (pp. 40–3). In very many instances the absolute dimensions confirm that the morphologist has made the most appropriate choice—the selected parametral plane proves in fact to be the 111 plane of the structural unit cell. Thus for aragonite an X-ray determination gave: $a_0 = 4\cdot94$, $b_0 = 7\cdot94$, $c_0 = 5\cdot72$ A., which are in the ratios $0\cdot6222 : 1 : 0\cdot7204$; the axial ratios quoted by Dana are $a : b : c = 0\cdot6224 : 1 : 0\cdot7206$. Occasionally, however, the conventionally chosen parametral plane is not the structural 111 plane but is found to correspond to the use of a simple multiple of the unit dimension along one, or two, of the axes (see Fig. 59, p. 42). For barium sulphate (the mineral barytes) morphological reference books quote $a : b : c =$

0·8152 : 1 : 1·3136; the unit cell dimensions are $a_0 = 8·8625$, $b_0 = 5·4412$, $c_0 = 7·1401$ A., whence $a_0 : b_0 : c_0 = 1·6288 : 1 . 1·3123$, and we can see that the selected parametral plane is structurally the plane 211 in terms of the unit cell.

We are now in a position also to determine the number Z of formula units associated with the unit of pattern (the 'contents of the unit cell'). The volume of the unit cell of barytes $= 8·86 \times 5·44 \times 7·14 \times 10^{-24}$ c.c., the molecular weight is 233, and the 'unit of atomic weight' $1·66 \times 10^{-24}$ gms. The density is 4·5, so that

$$\frac{233 \times 1·66 \times 10^{-24} \times Z}{8·86 \times 5·44 \times 7·14 \times 10^{-24}} = 4·5$$

whence $Z = 4$. The unit of pattern therefore contains four barium ions and four SO_4 groups. Before we can determine the actual position of these within the unit cell the crystal must be assigned to its correct space group, a step which involves *indexing* the spots on the photographs.

In the z-axis rotation photograph of an orthorhombic crystal considered above it was seen that spots on the equator line must be due to reflections from families of planes $h\ k\ 0$. When the cell dimensions a_0 and b_0 have been determined from two further photographs, the spacings d_{hk0} of such planes can be calculated, and substitution in the Bragg equation $n\lambda = 2d \sin \theta$ gives the appropriate values of the angle of reflection. By measurement of the distances x on the photograph (Fig. 494) the θ-values for equatorial spots actually present are calculated from the relationship

$$\frac{x}{r} = 2\theta \text{ radians}$$

and comparison of the two lists enables us to assign the correct indices to each spot. Spots on the layer lines could be dealt with in a similar manner, by measuring x and y coordinates on the film, though the geometry here is rather more intricate. Certain regularities on t1e film are of some assistance, too. Indices on the first layer lines of this particular photograph will all be of the type $h\ k\ 1$, and in general on the nth layer lines of the type $h\ k\ n$; successive spots $h\ k\ 0$, $h\ k\ 1$, $h\ k\ 2 \ldots h\ k\ n$ lie above each other on *row lines*, the shape of which can be calculated and which, near the centre of the photograph, approximate to vertical lines. This *ad hoc* approach to the problem of indexing, however, becomes more and more involved as the symmetry of the crystal declines towards triclinic, when even the calculation of a list of spacings for comparison is decidedly tedious, and it has been completely superseded by graphical procedures into which it is not our purpose to enter here. The reader who wishes to pursue the matter further is recommended to consult *The Interpretation of X-ray Diffraction Photographs* by N. F. M. Henry, H. Lipson and W. A. Wooster. Modifications have been introduced also into the practical technique of the simple rotation photograph in order further to simplify the interpretation and to remove certain possible ambiguities. In place of a complete revolution, the crystal is more usually *oscillated* through a small angle (5°–15°); frequently the film, too, is moved continuously during the exposure, as in the *Weissenberg* apparatus in which a translation of the cylindrical camera parallel to its axis is coupled to the rotation of the crystal. In some of these methods an important contrast with the Laue procedure is the lack of direct correlation between the symmetry of the photograph and the symmetry of the crystal structure about the direction of the incident beam. For example, whilst the rotation photograph of an orthorhombic crystal described above will be symmetrical about the equator line and about the vertical, the usual type of oscillation photograph of the same crystal would display only equatorial symmetry.

The photographic procedures so far described have in common the necessity for the material under investigation to be such that a single crystal or fragment of a single crystal can be isolated (it is of course unnecessary for the material to show external crystal faces, this being an advantage only when setting up the crystal in the desired orientation on the apparatus). To complete our outline of available methods one further type of photograph must be described. If the material is found only in the form of a fine powder or a fine-grained disoriented aggre-

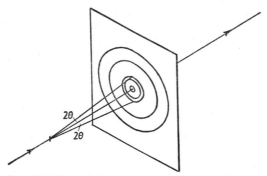

Fig. 497. Diagram of a powder camera using a flat plate.

gate, a *powder photograph* is the only type of X-ray photograph which can be obtained. When a sample of the powder is placed in the path of an X-ray beam diffracted X-rays will lie on a series of cones coaxial with the path of the incident beam (Fig. 497). If these diffracted rays are received on a flat plate, as in some of the earlier types of powder camera, the powder pattern will appear on development as a series of concentric dark rings. It is unnecessary, however, to record more than a portion of each ring, and moreover a flat plate will not receive rays deviated through large angles; hence in modern cameras the photograph is taken on a narrow strip of film laid around the inside of a squat cylindrical camera, on the axis of which the specimen is mounted (Fig. 498). A really fine powder presents in the path of the beam

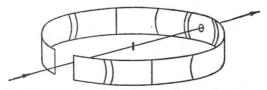

FIG. 498. Diagram of a cylindrical camera for powder photography.

particles with an infinite variety of orientations in space, yielding continuous dark lines on the photograph; a coarse sample tends to give lines of spots owing to the absence of grains of certain orientations. Since such spotty lines are difficult to measure accurately the sample is usually rotated during the exposure. When the film is laid flat after development, the powder pattern appears as shown diagrammatically in Fig. 499.

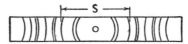

FIG. 499. Appearance of a powder photograph when the film is laid flat.

With a substance of cubic symmetry the photograph can be readily interpreted. If the distance S (Fig. 499) between corresponding lines on either side of the direct beam is measured on the film, the appropriate θ-value for the set of reflecting planes in question is given by the relationship $\theta = \dfrac{S}{4\,r}$ where r is the radius of the camera. The spacing of a set of planes hkl in the cubic system is

$$d_{hkl} = \frac{a_0}{\sqrt{h^2 + k^2 + l^2}} \quad \dots\dots\dots\dots\dots\dots\dots\dots\dots\dots\dots\dots\dots\dots\dots\dots(1)$$

where a_0 is the length of edge of the cubic unit cell, and inserting this value in the equation $n\lambda = 2d \sin \theta$ we have

$$\sin^2 \theta = \frac{n^2\lambda^2}{4a_0^2} (h^2 + k^2 + l^2) \quad \dots\dots\dots\dots\dots\dots\dots\dots\dots\dots\dots(2)$$

Hence, *the squares of the sines of the angles of reflection are proportional to the sums of the squares of the indices of the reflecting planes.*

The table sets out the calculations for a photograph of sodium chlorate taken in a camera of radius 5·0 cms. with radiation $\lambda = 0·709$ A.; the values in the column headed $q \times (h^2 + k^2 + l^2)$ can be written down by inspection after it has been observed that the ratios of the first four $\sin^2 \theta$ values are approximately $1 : 1·5 : 2 : 2·5$.

TABLE I

Powder photograph of sodium chlorate

S	θ rad:	$\theta°$	$\sin^2\theta$	$q \times (h^2+k^2+l^2)$	hkl
1·53	0·0765	4° 23′	0·00584	0·00292 × 2	110
1·87	·0935	5° 21′	·00870	·00290 × 3	111
2·16	·1080	6° 11′	·01160	·00290 × 4	200
2·41	·1205	6° 54′	·01442	·00288 × 5	210
2·65	·1325	7° 36′	·01750	·00292 × 6	211
3·07	·1535	8° 48′	·02341	·00293 × 8	220
3·26	·1630	9° 20′	·02631	·00292 × 9	300, 221
3·44	·1720	9° 51′	·02927	·00293 × 10	310
3·60	·1800	10° 19′	·03207	·00292 × 11	311
3·78	·1890	10° 50′	·03534	·00295 × 12	222
3·90	·1950	11° 10′	·03748	·00288 × 13	320
4·06	·2030	11° 38′	·04069	·00291 × 14	321
4·36	·2180	12° 29′	·04670	·00292 × 16	400
4·50	·2250	12° 53′	·04968	·00292 × 17	410, 322
4·61	·2305	13° 12′	·05216	·00290 × 18	411, 330
4·74	·2370	13° 35′	·05509	·00290 × 19	331
4·87	·2435	13° 57′	·05808	·00290 × 20	420
5·01	·2505	14° 21′	·06140	·00292 × 21	421
5·12	·2560	14° 40′	·06411	·00291 × 22	332

Calculation of cell size.

Average value of $q = 0·00291$ $\quad a_0^2 = \dfrac{(0·709)^2}{4 \times 0·00291}$

whence $\qquad\qquad\qquad\qquad a_0 = 6·57$ A.

Two points concerning the indices listed in the column *hkl* must be noted. In the first place we have been accustomed so far to reducing crystallographic indices to their simplest terms but in this column multiple indices, as 200, 220, 400 etc., appear. This is the X-ray crystallographer's device for indicating the *order* of reflection in question—220 denotes a second order reflection from planes 110, whilst 200 and 400 denote respectively the second order and the fourth order reflections from planes 100. (From the Bragg equation it is readily seen that the value of θ for the nth order reflection from planes of actual spacing d is the same as that for a first order reflection from a set of planes with spacing d/n, and it is this spacing which would be determined by using the appropriate multiple indices when calculating relationships such as that given for a cubic crystal by equation (1) above). Secondly, alternative values of $h\,k\,l$ are given for certain values of $h^2+k^2+l^2$, as 300 and 221, 410 and 322 etc. It is a defect of the powder method that, since the position of a line on the film depends only on the θ-value, we cannot distinguish between the contributions made to a particular line by two or more sets of different planes for which the θ-values are identical. Tables for X-ray crystallography list all possible values of $h\,k\,l$ to a reasonable limiting value of $h^2+k^2+l^2$ and also tabulate values of $\sin^2\theta$. The powder photograph having been indexed, the length of the edge a_0 of the unit cell can be calculated by substituting in equation (2) corresponding values of $\sin^2\theta$ and $h\,k\,l$; the most accurate results are obtained by using lines of high θ-value, or an average can be calculated by first determining the average value of the constant q.

Ingenious graphical methods have been evolved by which the interpretation of powder photographs can be effected directly without calculation, and these methods are of some assistance in the more difficult task of interpretation in systems other than cubic. The difficulties, however, increase rapidly with declining symmetry so that, whilst powder photographs have many exceedingly important applications in other fields of X-ray crystallography, they are ill-suited in general for use in determinations of cell-size and crystal structure and we shall not pursue the subject further here. We tabulate below two further examples of calculations for cubic crystals to which we shall have occasion to refer later. In the calculation for chromium it may be noted that the first four values of $\sin^2\theta$ are in the ratios of $1:2:3:4$ and we might have set out to use a value of q of approximately 0.030; this, however, would lead to a value $h^2+k^2+l^2=7$ for the seventh line and since this is an impossibility the true value of q must be 0.015 and the first line must correspond to $h^2+k^2+l^2=2$.

TABLE II

Powder photograph of periclase

S	θ rad:	$\theta°$	$\sin^2\theta$	$q \times (h^2+k^2+l^2)$	hkl
2·94	0·1470	8° 25′	0·02149	0·007163 × 3	111
3·39	·1695	9° 43′	·02848	·007120 × 4	200
4·81	·2405	13° 47′	·05673	·007091 × 8	220
5·66	·2830	16° 13′	·07800	·007091 × 11	311
5·95	·2975	17° 3′	·08597	·007164 × 12	222
6·89	·3445	19° 44′	·1140	·007125 × 16	400
7·56	·3780	21° 40′	·1363	·007174 × 19	331
7·74	·3870	22° 10′	·1423	·007115 × 20	420
8·50	·4250	24° 21′	·1699	·007079 × 24	422

Calculation of cell size.

$$\text{Average value of } q = 0.007125 \quad a_0^2 = \frac{(0.709)^2}{4 \times 0.007125}$$

whence
$$a_0 = 4.20 \text{ A.}$$

Powder photograph of chromium

S	θ rad:	$\theta°$	$\sin^2\theta$	$q \times (h^2+k^2+l^2)$	hkl
3·49	0·1745	10° 0′	0·03014	0·01507 × 2	110
4·97	·2485	14° 14′	·06047	·01512 × 4	200
6·16	·3080	17° 39′	·09193	·01532 × 6	211
7·17	·3585	20° 32′	·1230	·01537 × 8	220
8·01	·4005	22° 57′	·1520	·01520 × 10	310
8·82	·4410	25° 16′	·1823	·01519 × 12	222
9·57	·4785	27° 25′	·2121	·01515 × 14	321
10·30	·5150	29° 30′	·2425	·01516 × 16	400
10·97	·5485	31° 26′	·2719	·01510 × 18	411
11·74	·5870	33° 38′	·3068	·01534 × 20	420
12·40	·6200	35° 31′	·3375	·01534 × 22	332
12·89	·6445	36° 56′	·3611	·01504 × 24	422

Calculation of cell size.

$$\text{Average value of } q = 0.01520 \quad a_0^2 = \frac{(0.709)^2}{4 \times 0.01520}$$

whence
$$a_0 = 2.87 \text{ A.}$$

DETERMINATION OF SPACE GROUP

Accepting, then, that with a fuller understanding we could readily complete the indexing of a series of photographs of a given substance, the next step towards the complete determination of the crystal structure is to allocate the crystal to the appropriate space group. Here use is made of *systematic absences* from the lists of indices of reflections actually recorded on the photographs. We have seen that the 1st, 2nd, 3rd, . . . nth orders of reflection, from a series of planes with spacing d, will occur at angles

$$\sin^{-1}\frac{\lambda}{2d} \quad \sin^{-1}\frac{\lambda}{d} \quad \sin^{-1}\frac{3\lambda}{2d} \cdot \cdot \cdot \cdot \cdot \sin^{-1}\frac{n\lambda}{2d}.$$

If, however, these planes are regularly interleaved by a similar set, so that the actual spacing is reduced to $d/2$, then the reflections will occur only at angles

$$\sin^{-1}\frac{\lambda}{d} \quad \sin^{-1}\frac{2\lambda}{d} \quad \sin^{-1}\frac{3\lambda}{d} \dots \sin^{-1}\frac{n\lambda}{d},$$

corresponding to the *even* orders only from the first list, the odd orders now being systematically absent.

Many of the translations involved in building up the space groups (Chaps. X and XI) result in such regular interleaving of structural

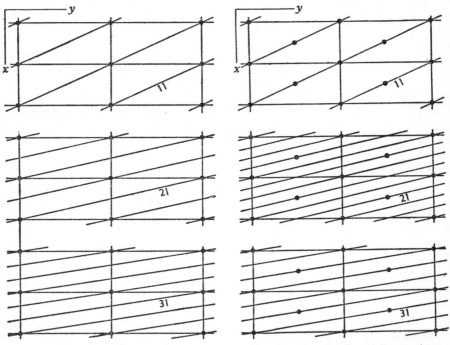

FIG. 500. Spacings of some *hk* lines in a primitive rectangular network.

FIG. 501. Corresponding *hk* lines to those of Fig. 500 is a centred network.

planes.   Considering first for simplicity a two-dimensional pattern, Fig. 500 shows graphically the spacing of some simple *h k* lines in a primitive rectangular network.   In Fig. 501 corresponding *h k* lines are drawn in a centred network, and it will be readily seen that the spacing is *halved* unless $h+k=2n$.   Extending this conception to three dimensions, Fig. 501 would represent the projection on 001 of a *C* face-centred lattice, and since equivalent atoms are associated with *every* point in such a lattice we can deduce that reflections of general type $h\,k\,l$ will only appear if $h+k$ is even. Similarly, an A face-centred lattice would be indicated by the absence of $h\,k\,l$ reflections unless $k+l$ is even, whilst for a B face-centred lattice only reflections for which $h+l$ is even would be present. It can be readily deduced that for an all face-centred lattice F only those reflections $h\,k\,l$ for which h, k and l are all odd or all even (homogeneous indices) will appear; for a body-centred lattice the criterion is that reflections are absent unless $h+k+l$ $=2n$. Only if we can be sure that none of these restrictions applies to the list of indices derived from our photographs are we justified in assuming that the structure is based on a primitive lattice P. Looking back at the list of indices derived from the powder photograph of periclase (p. 285) it is clear that they correspond to an F lattice, ' mixed ' indices being characteristically absent. Chromium, showing only indices such that $h+k+l=2n$, is based on an I lattice. In the list for sodium chlorate (p. 283) all possible types appear without restriction (the nineteen lines indexed correspond, apart from the absence of a 100 reflection, to successive possible values of $h^2+k^2+l^2$) and the lattice must be primitive.

Systematic absences from the $h\,k\,l$ reflections thus indicate that we are using a multiply primitive cell (p. 227). There are other operations of translation within the space groups, however, which result in systematic absences of a more restricted kind. In Fig. 502 is shown a

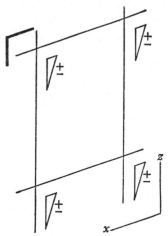

Fig. 502. A portion of a pattern based on the space group *Pm*.

portion of a pattern based on the space group *Pm* (viewed along the crystallographic *y* axis), and in Fig. 503 a related pattern based on the

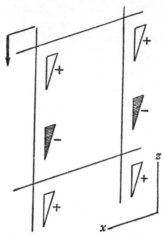

FIG. 503. A portion of a pattern based on the space group *Pc*.

space group *Pc*. The operation of the *c* glide plane clearly introduces possibilities of halving and resultant characteristic absences. Since, however, the representative units are here alternately above and below the paper, only planes *h* 0 *l*, *normal to the* 010 *glide plane*, have their spacing systematically *halved*; for such planes, *l* must be even for a reflection to be recorded. Reflections from general planes *h k l* will be affected in *intensity*, in comparison with those for the group *Pm*, but they will not show any *systematic* absences. A diagonal glide plane is indicated by the systematic absence of reflections unless the *sum* of the appropriate indices is even—for an *n* glide plane parallel to 100 $k+l=2n$ in all 0 *k l* reflections. If 100 is a *d* glide plane, 0 *k l* reflections appear only if $k+l$ is divisible by 4. The criteria for pinacoidal glide planes in the appropriate space groups may be summarised in the table opposite.

Glide planes in other than pinacoidal positions, such as the 110 glide planes in some tetragonal and cubic space groups, similarly give rise to systematic absences from reflections from planes at right angles to them. The space groups of periclase and of chromium (p. 285) are respectively *Fm3m* and *Im3m*, so that the lists of indices which we derived from the powder photographs show no systematic absences due to glide planes. The structure of spinel is based on the space group *Fd3m*, and the

Glide Plane	Type	Criterion for reflection
100	b glide c „ n „ d „	$0\,k\,l$ must have $k=2n$ $0\,k\,l$ „ „ $l=2n$ $0\,k\,l$ „ „ $k+l=2n$ $0\,k\,l$ „ „ $k+l=4n$
010	c „ a „ n „ d „	$h\,0\,l$ „ „ $l=2n$ $h\,0\,l$ „ „ $h=2n$ $h\,0\,l$ „ „ $h+l=2n$ $h\,0\,l$ „ „ $h+l=4n$
001	a „ b „ n „ d „	$h\,k\,0$ „ „ $h=2n$ $h\,k\,0$ „ „ $k=2n$ $h\,k\,0$ „ „ $h+k=2n$ $h\,k\,0$ „ „ $h+k=4n$

following list of indices was derived from a powder photograph: 111, 220, 311, 222, 400, 422, 333 and 511, 440, 531, 620, 533, 444, 551 and 711, 642, 553 and 731, 800, 660 and 822, 555 and 751, 840 . . . from which we can see the restriction $h+k=4n$ imposed by the presence of the d glide planes (in addition to restrictions imposed by the F lattice) on reflections of type $h\,k\,0$. In the highly symmetrical group $Ia3d$ (space group No. 230 of most lists) the lattice restriction imposes the condition $h+k+l=2n$ on all reflections $h\,k\,l$; as a consequence, l must be even in all reflections $h\,h\,l$ and moreover these are further restricted to $2h+l=4n$ by the presence of the d glide planes; and finally in all reflections $h\,k\,0$, h (and consequently k) must be even. A rotation photograph of a garnet, with a structure based on this space group, gave the following indices for the first five layer lines:

Zero line (Equator), 400, 420, 440, 620, 640, 800, 840, 10.40, 880, 12.60, 14.40.
First line, 321, 431, 521, 611.
Second line, 042, 422, 512, 532, 642, 842, 10.42, 12.22, 12.82.
Third line, 323, 413, 523.
Fourth line, 024, 224, 314, 044, 444, 064, 264, 084, 284, 664, 864 and 10.04.

Systematic absences of still more restricted range are caused by the operation of screw axes, which effect regular interleaving of structural planes to which they are normal, but not of any other planes (Fig. 509 p. 301). Space groups $P2$ and $P2_1$ would be distinguished by the fact that from a structure based on the latter the 0 k 0 reflections would be systematically absent unless k is even; for $P2_12_12_1$, whilst $h\,k\,l$ reflections would be present in all orders, the pinacoidal reflections $h\,0\,0$, $0\,k\,0$ and $0\,0\,l$ would be represented only by even indices. The absence of a 100 reflection from the powder photograph of sodium chlorate noted on p. 283 is in fact a consequence of the structure being based on the space group $P2_13$, but of course a more comprehensive range of reflections than these first few lines of a powder photograph must be studied before we can with certainty make deductions concerning systematic absences.

Extending this argument to other types of screw axes (pp. 228–30), we conclude that vertical axes 4_2 or 6_3 would likewise produce systematic halving of the spacing of basal pinacoids, with corresponding systematic absences; 3_1, 3_2, 6_2 and 6_4 reduce the spacing to one third; 4_1 and 4_3 cause systematic quartering; whilst for 6_1 and 6_5 only reflections $0\,0\,0\,l$ for which $l = 6n$ can be present.

One further piece of information is generally necessary before we can make a unique choice of space group. Systematic absences arise from the action of symmetry operations involving *translation* but not from those of *reflection* planes or of *rotation* axes. Laue photographs of appropriate orientation will classify the crystal in one of eleven groups (p. 156), but X-ray methods cannot in themselves give *direct* information on the presence or absence of a centre of symmetry. (Under all ordinary conditions of working, reflections $h\,k\,l$ and $\bar{h}\,\bar{k}\,\bar{l}$, are identical; the statement that every crystal diffracts X-rays as if a centre of symmetry were present is sometimes known as *Friedel's Law*.) They may do this indirectly by proving the presence of a combination of symmetry elements which is necessarily centrosymmetrical; thus if systematic absences from photographs of a monoclinic substance indicate the presence both of a screw axis 2_1 and a glide plane c then the point group must be $2/m$ and the space group is uniquely determined as $P2_1/c$, and similarly the presence of three pinacoidal glide planes in an orthorhombic crystal, such as in the space groups $Pcca$ or $Fddd$, must mean that the crystal belongs to the holosymmetric point group mmm. This applies, however, to only 42 of the space groups; for the remainder, the final allocation of a crystal to a single point group must still be accomplished by reference to external morphology or by physical investigation (pp. 151–5), and study of the systematically absent reflections then determines the space group uniquely.

Thus, for example, photographs of the monoclinic mineral pickeringite showed no systematic absences from a wide range of recorded indices, indicating that the lattice is primitive and that no glide plane or screw axis is present in the space group. On the assumption that it belonged to the holosymmetric class of the monoclinic system the space group was hence given as $P2/m$. Later work on the morphology assigned it quite clearly to class 2 (p. 111), so that the correct space group must be $P2$. In a study of marcasite, orthorhombic FeS_2, no systematic absences from $h\,k\,l$ reflections were observed, indicating a primitive lattice P. Reflections $h\,k\,0$ were systematically absent except for values $h+k=2n$, and reflections $h\,0\,l$ except when $h+l=2n$. Hence we deduce glide planes n parallel to 001 and 010. Reflections $0\,k\,l$ of all types were present, so that there are no glide planes parallel to 100, but the final allocation to the space group $Pmnn$ (conventional orientation $Pnnm$) rests on the knowledge that marcasite belongs to the holosymmetric class, mmm. If there were evidence that it lacked a centre of symmetry, the space group would be $Pnn2$ of the class $mm2$.

Yet one other point must be noted. Systematic absence of a more general type of reflection necessarily involves also systematic absence of related special types, and these further absences *do not imply* the presence of additional symmetry elements. Thus the photographs of marcasite show also the presence of reflections $h\,0\,0$ only when $h=2n$, reflections $0\,k\,0$ only when $k=2n$, and reflections $0\,0\,l$ only when $l=2n$, but these are merely special cases of the systematic absence of $h\,k\,0$ and $h\,0\,l$ reflections ; we cannot deduce from them the presence of screw axes, and they would be systematically absent also from photographs of a crystal built on the group $Pnn2$. The space groups $I222$ and $I2_12_12_1$ (p. 249) cannot be differentiated, since for both the characteristic absences are the same, and the same applies to the cubic groups $I23$ and $I2_13$.

DETERMINATION OF ATOMIC POSITIONS

The space group having been determined, the last step in a complete structural analysis involves the locating of the ' contents of the unit cell ' (p. 280) within this scaffolding of symmetry elements. With simple structures some help will usually be available from a study of the number of equivalent positions (p. 252) within the unit cell. Thus fluorite, CaF_2, has four formula-units per unit cell, space group $Fm3m$; the *International Tables* list the equivalent positions:

$$(000; 0\tfrac{1}{2}\tfrac{1}{2}, \tfrac{1}{2}0\tfrac{1}{2}; \tfrac{1}{2}\tfrac{1}{2}0) +$$
$$4: (a)\ 000.\quad (b)\ \tfrac{111}{222}.$$
$$8: (c)\ \tfrac{111}{444}; \tfrac{333}{444}.$$

Since there is only one set of eightfold equivalent positions, the eight F ions are placed there; the alternative fourfold equivalent positions

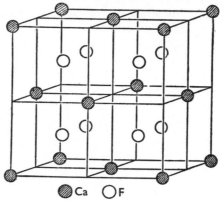

FIG. 504. Diagram of a portion of the structure of Fluorite. CaF₂.

for Ca then correspond merely to a change of origin, so that the structure is uniquely determined (Fig. 504).

Arsenides of the type RAs_3, such as the mineral skutterudite $CoAs_3$ contain 8 formula units per unit cell, space group $Im3$; the eightfold positions are:

$$\tfrac{111}{444}; \tfrac{133}{444}; \tfrac{313}{444}; \tfrac{331}{444};$$
$$\tfrac{333}{444}; \tfrac{311}{444}; \tfrac{131}{444}; \tfrac{113}{444};$$

and here we place the eight metal atoms. If all the arsenic atoms occupy one set of equivalent positions, they must be in the 24-fold special equivalent positions given by the co-ordinates:

$$(000; \tfrac{111}{222}) +$$
$$24: (g)\ 0yz;\ z0y;\ yz0;\ 0\bar{y}\bar{z};\ \bar{z}0\bar{y};\ \bar{y}\bar{z}0;$$
$$0y\bar{z};\ \bar{z}0y;\ y\bar{z}0;\ 0\bar{y}z;\ z0\bar{y};\ \bar{y}z0.$$

It must be borne in mind, however, that in many structures atoms of one kind may be divided between two or more types of equivalent position; the twelve 0 ions of aragonite (p. 253) are distributed between fourfold special positions of type $\tfrac{1}{4}yz$ and eightfold general positions of type $x_1 y_1 z_1$ (taking the origin at a centre of symmetry).

The final determination of the exact positions of atoms in all such cases involves the evaluation of one, two or three parameters x, y, z for each type of equivalent position. When determining the space group, attention was paid only to the *presence* or *absence* of particular reflections; now it becomes necessary to take into account the *intensities* of those reflections which *are* recorded. Referring again to a two-dimensional array for simplicity, there is represented in Fig. 505

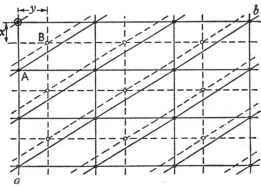

Fig. 505. A portion of a structural pattern, based on a rectangular network, containing atoms of two different kinds.

a structure in which atoms A of one kind, represented as black circles, are located at the corners 00 of the unit rectangle, whilst atoms B of another kind, represented by open rings, lie within the rectangle in a position with fractional co-ordinates xy. Families of lines, such as the 10, 01 and 11 lines illustrated, in such a structure consist of rows of atoms A interlined by rows of atoms B. The effect of such interlining is to modify the intensities of reflections from various rows in comparison with those which would result from the presence of A atoms alone. Supposing that atoms A and B had practically equal scattering powers for X-rays, and that x had a value of approximately $\frac{1}{2}$, then the effect would be to reduce the odd orders of reflections 10, almost to zero; if the value were approximately $\frac{1}{3}$ then it would be reflections 30, 60 etc. which would be particularly strong. In similar fashion, the ratio y/b_0 will determine the extent to which reflections 01 are modified in intensity by the presence of B atoms, in comparison with those which would arise from A atoms alone. The effect on 11 reflections depends upon the value of the ratio $\dfrac{a_0 - x}{b_0 - y}$ in comparison with the ratio $\dfrac{a_0}{b_0}$; if these are approximately equal, the B atoms will fall nearly upon the 11 rows of A atoms and an enhancement of reflections will result, but if the ratios are such that the 11 rows of B atoms interline the A rows almost exactly then the effect will be substantially to reduce some orders of reflection and to reinforce others.

The scattering power of atoms of different kind is proportional to their atomic number; knowing this, and taking into account also further factors including the angle of scattering and the wavelength of the X-rays, it is possible to calculate the intensities of reflections from a postulated arrangement. In choosing probable atomic positions, valuable help can be obtained from consideration of such features as the atomic (or ionic) sizes of the components, the presence of molecular groups, the effects of isomorphous substitution, the anisotropy of optical and other physical properties of the crystal, the presence or absence of cleavage directions and so forth. None the less, determination of parameters is ultimately essentially a process of trial and error, calculated intensities for a given arrangement being compared with observed intensities—either estimated visually from photographs, determined by photometry of the spots or measured directly by non-photographic methods. The problem of intensity measurements and their application, however, falls within the field of advanced crystallography and we must conclude that further discussion is beyond the scope of this *Introduction*.

The student may pursue the matter with the help of the following books :

BRAGG, W. L.   *The Crystalline State. Vol. 1.   A General Survey.* London, 1933.   A general account, very clearly written though inevitably now outdated in parts, which can be read with profit by any student.

BUNN, C. W.   *Chemical Crystallography.*   Oxford, 1945.   Chapters VI, VII and IX provide a clear description of methods of structure determination, with illustrative examples.

LONSDALE, K.   *Crystals and X-rays.*   London, 1948.   The author covers a wide field at a level demanding some background knowledge on the part of the reader.

BUERGER, M. J.   *X-ray Crystallography.   An Introduction to the Investigation of Crystals by Their Diffraction of Monochromatic X-Radiation.*   New York, 1942.   A thorough treatment within a somewhat specialized field.

CHAPTER XIII

CRYSTAL HABIT

THE RELATIONSHIP OF CRYSTAL HABIT TO THE STRUCTURAL PATTERN

We offer in this concluding chapter some observations on the relationships between the structural pattern and the external morphology of a crystal. Though a fuller understanding of these relationships has been achieved only in the last few years, on the basis of the results of investigations using X-rays, it will appear that it is in fact often possible to reach an unambiguous determination of the space group from a study of crystal habit alone.

This serves to emphasise the close relationship between the morphology and the details of the symmetry of internal arrangement. It must be understood, however, that such deductions are not used in practice to supersede the direct determination of a space group by the methods of X-ray crystallography which we have outlined in the preceding chapter; such a proceeding, indeed, would be most unsafe in the present state of our knowledge for we shall note at the conclusion of this chapter that in spite of growing understanding of the significance of external morphology there remain many anomalies still awaiting explanation.

Throughout our earlier discussions we have made frequent use of the concept that the planes which are most likely to appear as external faces are those most densely occupied by significant points of the unit of pattern. Fig. 506 shows in plan normal to [001] the traces of three sets of planes—100, 110, 210—from the [001] zone of a primitive orthorhombic arrangement. Of these three sets the 100 planes are most densely beset with significant points, followed in order by the 110 and the 210 planes. The more densely occupied the planes are, the more widely they are spaced—the *interplanar spacing d* of a given type of plane is directly proportional to the *reticular density* or inversely proportional to the *reticular area* S (the area of the smallest mesh in the net of points in the plane).

Fig. 507 illustrates similarly the corresponding sets of planes in a C-face-centred net, and it is clear that with the introduction of a doubly primitive unit of pattern some conditions have changed. The actual reticular area of the 100 planes is the same as that of the corresponding planes in Fig. 506, but the interplanar spacing is halved; the 110 planes, with the same spacing as in Fig. 506, are twice as densely

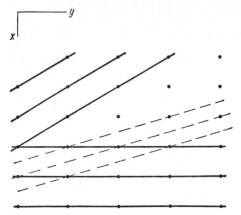

FIG. 506. Lattice planes in an orthorhombic *P* space lattice.

occupied; the 210 planes, like the 100 planes, show the same reticular area, with a spacing which is halved, when compared with the corresponding planes in the primitive pattern. Thus we might expect from

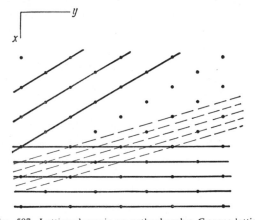

FIG. 507. Lattice planes in an orthorhombic *C* space lattice.

Fig. 507 that the order of decreasing relative morphological importance of the three forms would be {110}, {100}, {210}, in contrast with the order {100}, {110}, {210} derived from Fig. 506. Consideration of further sets of planes would show that in a pattern based on a C-face-centred lattice the interplanar spacing is halved (and hence the reticular area S is doubled) for all sets of planes $h\ k\ l$ for which $h+k$ is not even. With the all-face-centred unit of an F lattice the halving occurs for all planes in which any of the sums $h+k$, $k+l$, $l+h$ is not even—only planes with indices composed of three odd figures are unchanged in spacing in comparison with a pattern based on a P lattice. In a body-centred arrangement, only planes for which $h+k+l$ is even are unchanged in spacing.

If we restrict attention for the moment to the cubic system, the calculation of the relative spacings of different sets of planes is simple. If a_0 is the side of the unit cube of the pattern, the spacing of a set of planes $h\ k\ l$ in a primitive pattern is given by

$$d_{h\,k\,l} = \frac{a_0}{\sqrt{h^2 + k^2 + l^2}},$$

and hence in terms of S, the reticular area, $S^2_{h\,k\,l} \propto h^2 + k^2 + l^2$. Calculating thus for a few simple planes:

$h\ k\ l$	100	110	111	210	211	221	310	311	320	321	410	322	411
S^2	1	2	3	5	6	9	10	11	13	14	17		18

This order of increasing reticular areas we should expect to represent an order of decreasing morphological importance of the corresponding forms in cubic crystals based on a primitive lattice.

How can we carry out the corresponding calculations for a pattern based on an F lattice? We have seen that the spacing is halved, and S therefore doubled, for all planes for which the index hkl consists of mixed odd and even figures. To correct for this, we must double all such 'mixed' indices, and use the multiple index in the calculation of d or S.

$h\ k\ l$	200	220	111	420	422	442	620	311	640	642	331	511	531
S^2	4	8	3	20	24	36	40	11	52	56	19	27	35

Rewriting the forms in the order of importance indicated by increasing reticular area, we obtain the list
111, 100, 110, 311, 331, 210, 211, 511, 531, 221 ... ,
suggesting that cubic crystals based on an F lattice may be expected to

show the octahedron predominating over the cube (*octahedral mode*, contrasting with the *hexahedral mode* of those based on a primitive arrangement).

For a body-centred *I* lattice we must double any index for which the sum of the figures is not even.

h k l	200	110	222	420	211	442	310	622	640	321	411	332	431
S²	4	2	12	20	6	36	10	44	52	14	18	22	26

From these figures we derive an order of importance:

110, 100, 211, 310, 111, 321, 411, 210, 332, 431 ... .

Here the dodecahedron predominates, and the crystals can be described as based on a *dodecahedral mode*. We begin to see a possible explanation of the predominantly cubic habit of some substances, such as $NaClO_3$ and CsCl; of the morphological importance of the octahedron in fluorspar and diamond; of the usual dodecahedral habit of the garnets and sodalite.

Similar calculations can be carried out in the other crystal systems, the general formula reading:

$$S^2_{hkl} = h^2 . b^2 c^2 \sin^2\alpha + k^2 . c^2 a^2 \sin^2\beta + l^2 . a^2 b^2 \sin^2\gamma$$
$$+ 2hk . abc^2 \sin\alpha . \sin\beta . \cos\nu + 2kl . bca^2 \sin\beta . \sin\gamma . \cos\lambda$$
$$+ 2lh . cab^2 \sin\gamma . \sin\alpha . \cos\mu,$$

where α, β, γ are the axial angles and

$$\lambda = 010 \frown 001, \quad \mu = 001 \frown 100, \quad \nu = 100 \frown 010.$$

The calculations in the less symmetrical systems are somewhat tedious. Tables have been constructed to simplify the work, or we can alternatively use a simple graphical solution. Mallard showed in his *Traité de Cristallographie* that in the gnomonic projection the distance of any pole *h k* 1 from the centre *O* of the sphere of projection (Fig. 508) is a comparative measure of the reticular area of the corresponding planes; the horizontal distance, *D*, of the pole from the centre of the gnomonic projection is easily determined graphically from the gnomonogram, and $S^2 = D^2 + r^2$. For poles *h k l* the corresponding central distance is $1/l$ of the required value, so that in general

$$S^2_{hkl} = (D^2_{hkl} + r^2) l^2.$$

As an example from the orthorhombic system we may choose orthorhombic sulphur. Simple crystals (Fig. 123, p. 71) show the bipyramid

{111} predominant, with {113}, {001} and {011}, suggesting the orthorhombic ' octahedral ' mode and an F lattice. This suggestion is con-

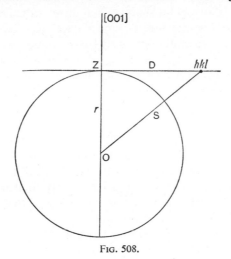

FIG. 508.

firmed by the kind of index characteristic of the forms developing on
more complex crystals. In Fig. 124, the additional forms present are
the pinacoids {100}, {010}; the prism {110}; domes {101}, {103},
{031} and {013}; bipyramids {112}, {113}, {114}, {115}, {117}, {221},
{331}, {311}, {313}, {315}, {131}, {133}, {135}. The striking feature in this
list is the predominance of planes with odd indices, precisely what we
should expect if the crystal is built on an F lattice. Calculation of the
values of S^2, doubling mixed indices in view of the lattice type, gives
the following relative values:

$h\,k\,l$	002	111	020	113	022	200	202	024	115	220	204	131	133
S^2/a^2b^2	4	10·1	14·5	18·1	18·5	21·9	25·9	30·5	34·1	36·4	37·9	39·1	47·1

$h\,k\,l$	026	224	311	206	117	313	042	135	315	028	240	331	242
S^2/a^2b^2	50·5	52·4	53·9	57·9	58·1	61·9	61·9	63·1	77·9	78·5	79·8	82·9	83·8

Thus of the 23 forms of the crystal of Fig. 124, all but 3 occur in the
first 25 forms listed in order of increasing reticular area.

A further important deduction which we may make from the general
formula for the calculation of S reveals that the relative dimensions of

the unit of pattern may also play a significant part in relation to the dominant habit. In a crystal with very unequal unit cell dimensions some planes of relatively high indices may have a smaller reticular area than others with simpler indices, and so we may expect the former rather complex forms to develop fairly frequently on specimens of such a substance.

Considerations of this kind were known to early crystallographers. They were discussed by Frankenheim and elaborated by Bravais (though not quite in the form in which we have developed them here). The statement that the forms which tend to occur most frequently on crystals are those with faces parallel to planes of smallest reticular area has consequently been termed the *Law of Bravais*. It may be noted that the Law of Rational Indices, as we developed it from considera- tion of Haüy's decrements, is implicit in this further law.   Bravais himself, and Mallard following him, developed the principle on a theoretical basis, and at first it received little attention outside France. The ultimate recognition of its importance was largely due to the work of another French crystallographer, G. Friedel,* who by publishing many convincing examples of its application (including the discussion of the forms of orthorhombic sulphur which we have used above) established it firmly as a fundamental law of observation.   We may conclude that there is convincing evidence of a relationship between the kind of structural pattern on which the crystal is based and the morphological importance of various forms.

The picture presented thus, in terms of the Bravais lattice only, is incomplete, however.   In the example of sulphur, though the corre- spondence in general is excellent, there are certain anomalies.   {001} at the head of the list suggests a predominance of the basal pinacoid, and hence probably a tabular habit, and the other pinacoids are of

---

* Georges Friedel was born at Mulhouse, Alsace, in 1865. The son of another mineralogist, C. Friedel, he inherited his father's interests, although his earlier life was spent in the Corps des Mines. In 1919 he became Professor of Mineralogy and Crystallography in the University of Strasbourg. His original work covered many fields in crystallography, mineralogy and geology, and he published text-books of mineralogy and of crystallography. He died at Strasbourg in 1933. The spirit of much of his work may be illustrated by some excellent advice offered in the preface to his *Leçons de Cristallographie* (1926): ' Trop souvent on enseigne aux jeunes gens . . . que lorsqu'ils veulent entreprendre une recherche, leur premier et d'abord unique soin doit être d'en réunir et d'en compulser la " Literatur ". Ce n'est que lorsqu'ils se seront farci la tête de tout ce qui a été écrit sur le sujet, et par consé- quent de dix erreurs pour une vérité . . . et auront ainsi perdu toute fraîcheur d'im- pression, qu'ils seront admis à regarder les faits par eux-mêmes. . . . Celle que nous préconisons est autre. La lecture, cela va de soi, y tient sa place, mais au second rang. L'essentiel, avant de lire, est de se mettre en face des faits, d'observer, expérimenter et réfléchir sans subir *a priori* l'influence de ce qu'ont pu dire *X* ou *Y*.'

minor morphological importance compared with their positions third and sixth in the list. The eighth form in order of increasing reticular areas, {012}, and the eleventh, {102}, are essentially unknown on sulphur crystals. The correspondence, in fact, is really far from perfect.

If we return for a moment to the simpler case of the cubic system we are faced again by the incompleteness of the explanation so far offered. We have recognised hexahedral, octahedral and dodeca-hedral modes, but how are we to account for a habit such as that typical of the garnets, in which the icositetrahedron {211} is often developed alone, or with {110} subordinate? Even in the list for an I lattice the form {211} is preceded by {110} and {100}. In the develop-ment of the space groups we recognised that a rotation axis in the external symmetry may be paralleled by screw axes in some of the isomorphous space groups; if the spacing of a given set of planes normal to a rotation diad axis 2 be d, the effect of the presence of screw diad axes 2_1 is to reduce the spacing to $d/2$ (Fig. 509). Thus the

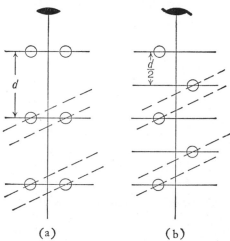

FIG. 509. The effect of an axis 2_1 on the spacing of planes normal to it; the arrangement of planes inclined to the axis is the same in both (a) and (b).

reticular area, inversely proportional to the spacing, is in effect doubled, and the corresponding form will recede in morphological importance to a place in the list appropriate to the doubled S value. Notice, how-ever, that this effect applies only to planes normal to the axis, for in directions inclined to the axis there do not arise new equidistant structurally equivalent planes (Fig. 509) as a consequence of the screw character of the axis. Whilst in considering the effect of a particular lattice type we must work with general indices $h\,k\,l$ (including also, of course, special planes such as $h\,k\,0$ and $h\,0\,0$), the presence of a screw axis affects *only* planes normal to the axis.

A screw triad axis 3_1 or 3_2 will reduce the spacing of planes normal to it by a factor $1/3$ (Fig. 510), so that the effective reticular area will be multiplied by 3 in comparison with a structure based on a space group containing only rotation triad axes. The Bravais principle applied to the study of quartz places the basal pinacoid $\{0001\}$ first in order of expected morphological importance, but the expectation that screw triad axes are present in the appropriate space group justifies calcula-

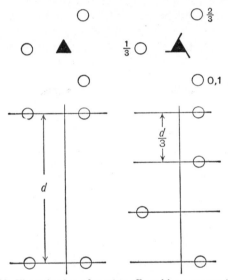

Fig. 510. The reduction of spacing effected by a screw triad axis.

tion with $\{0003\}$, and the base recedes to a position in accordance with its rarity as a growth form on quartz crystals. We may summarise the effects of all the types of screw axes which we have encountered:

Axes 2_1 or 4_2 or 6_3 multiply the S values of planes normal to them by 2.

,,	3_1 or 3_2	,,	,,	,,	,,	,,	,,	3.
,,	4_1 or 4_3	,,	,,	,,	,,	,,	,,	4.
,,	6_1 or 6_5	,,	,,	,,	,,	,,	,,	6.
,,	6_2 or 6_4	,,	,,	,,	,,	,,	,,	3.

In the cubic system, for example, the order of morphological importance for structures based on the space group $P23$ or $P432$ will be the
one which we calculated from consideration of the Bravais lattice alone
(p. 275):

$$100, \ 110, \ 111, \ 210, \ 211 \ \dots .$$

In the groups $P2_13$ and $P4_232$, however, we must calculate with the
index 200, and the cube recedes in importance to third place:

$$110, \ 111, \ 100, \ 210, \ 211 \ \dots ,$$

whilst in the groups $P4_132$ and $P4_332$ the value $S^2_{100} = 16$, and the cube
has fallen still further to the tenth position:

$$110, \ 111, \ 210, \ 211, \ 321 \ \dots .$$

Extending this argument we may examine next the effect which the
presence of glide planes in a space group may bring about in the calculation of the appropriate reticular areas. Fig. 511 shows in plan a

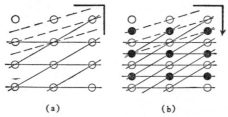

(a)   (b)

Fig. 511. In (a) the horizontal planes of symmetry are m planes, and the reflections of the open rings are vertically above and below them. In (b) the black
circles arise from the open rings by reflection in a glide planes.

portion of an orthorhombic P pattern with reflection planes parallel
to 001 and the corresponding arrangement if the 001 planes in the
space group are a glide planes. The new points, shown black, introduced by the glide planes result in the spacing of certain planes in the
zone [001] being halved. Of those drawn in the figure, 100 and 110
planes are 'halved', but the spacing of 210 planes, in which h is even,
is the same in both diagrams. In calculating S values for the second
arrangement we should use the doubled indices 200 and 220; or, in
general, double any index $h \, k \, 0$ in which h is odd. As with screw axes,
this effect of glide planes is specialised, applying only to the zone of
planes normal to the glide plane (compare the full and broken planes
in Fig. 512), and certain planes are thus decreased in morphological
significance. Diagonal glide planes n parallel to 001 are accounted for

by calculating with indices $h\,k\,0$ modified, if necessary, so that $h+k$ is even (Fig. 513), whilst if d planes are present the indices $h\,k\,0$ used in calculating must all be of the type with $h+k$ a multiple of 4.

Planes of this last type are present in the accepted structural arrangement of orthorhombic sulphur, for which the space group is *Fddd*, and their presence will modify the order of expected morphological importance deduced from consideration of the lattice type alone. Instead of 002, 020, 200 we must calculate with 004, 040 and 400, since these

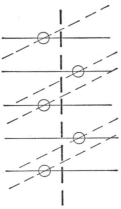

Fig. 512. The full lines represent in cross section *any* family of lattice planes normal to the glide plane (not merely a set of planes normal to the paper). The spacing of all such planes is halved by the action of the glide plane.

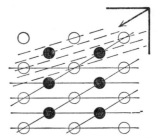

Fig. 513. The black circles arise from the open rings by reflection in horizontal *n* planes.

pinacoids are normal to d planes. The form {001} becomes second in importance to the prominent bipyramid {111}, and the morphologically unimportant pinacoids {010} and {100} recede down the list. Similarly, 012 must be calculated as 048 and 102 as 408, with the result that these forms also recede to positions of little morphological significance; the first form unobserved on sulphur has dropped from the 8th to the 20th place in the list.

$h\,k\,l$	004	111	040	113	022	400	202	048	115	220	408	131	133
S^2/a^2b^2	(16)	10·1	(58)	18·1	18·5	(87·6)	25·9	(122)	34·1	36·4	(152)	39·1	47·1

$h\,k\,l$	026	224	311	206	117	313	084	135	315	04·16	480	331	242
S^2/a^2b^2	50·5	52·4	53·9	57·9	58·1	61·9	(248)	63·1	77·9	(314)	(319)	82·9	83·8

Anhydrous sodium sulphate, Na_2SO_4, referred to the same space group, shows interesting analogies with orthorhombic sulphur in its morphological development, although the different values of the axial parameters, of course, introduce some variations in detail.

This important generalisation of the Bravais principle by consideration of the space group rather than the lattice type only was first fully developed in 1937 by two crystallographers working at the Johns Hopkins University, Baltimore—J. D. H. Donnay and D. Harker, (it had been treated briefly by P. Niggli of Zürich twenty years earlier). Whatever the possibility of establishing a theoretical justification, we must extend to the wider *Donnay-Harker Principle* the same recognition as a law of observation which was formerly given to the narrower Bravais Principle. The latter, of course, is included in the former, since the deductions from the two are identical for crystals based on those space groups containing neither screw axes nor glide planes.

We proceed next to a more detailed examination of some of the implications of this principle, once more turning first to the cubic system with its greater regularity. The Bravais Principle enabled us to establish three different series for decreasing morphological significance of various forms corresponding to the hexahedral, octahedral and dodecahedral modes. Such a series is termed by Donnay and Harker a *morphological aspect*, and we have already examined the effect of the presence of screw axes in modifying the morphological aspect appropriate to cubic crystals based on a primitive lattice. Similarly, the presence of n glide planes parallel to the cube faces (as in $Pn3$ and $Pn3m$) will decrease the importance of $\{h\,k\,0\}$ forms unless $h+k$ is even. Glide planes n in the dodecahedral directions ($P\bar{4}3n$, $Pm3n$, $Pn3n$) will affect $\{h\,h\,l\}$ forms unless l is even. In structures based on an F lattice or an I lattice, d planes may be present and will reduce still further the importance of certain $\{h\,k\,0\}$ forms (in $Fd3$, $Fd3m$ and $Fd3c$) or of certain $\{h\,h\,l\}$ forms (in $I\bar{4}3d$ and $Ia3d$). In this way the three morphological aspects appropriate to the cubic system on the principle enunciated by Bravais give place to seventeen different aspects under the generalised principle. We shall not attempt to derive them all here; the table below (which the student can check on his own account) lists the first five forms in order of decreasing importance for each aspect. (Where the first five forms are identical, differences would appear amongst the less important forms.)

	$P23$	$Pm3$	$P\bar{4}3m$	$P432$	$Pm3m$	100	110	111	210	211
	$P2_13$			$P4_232$		110	111	100	210	211
				$P4_132,P4_332$		110	111	210	211	221
P		$Pn3$			$Pn3m$	110	111	100	211	221
			$P\bar{4}3n$		$Pm3n$	110	100	210	211	310
					$Pn3n$	110	100	211	310	111
		$Pa3$				111	100	210	211	110
	$F23$	$Fm3$	$F\bar{4}3m$	$F432$	$Fm3m$	111	100	110	311	331
				$F4_132$		111	110	311	100	331
F		$Fd3$			$Fd3m$	111	110	311	100	331
			$F\bar{4}3c$		$Fm3c$	100	110	111	210	211
					$Fd3c$	110	111	100	211	531
	$I23, I2_13$	$Im3$	$I\bar{4}3m$	$I432$	$Im3m$	110	100	211	310	111
				$I4_132$		110	211	310	111	321
I		$Ia3$				100	211	110	111	321
			$I\bar{4}3d$			211	110	310	321	100
					$Ia3d$	211	110	321	100	210

Some important deductions can be drawn from a consideration of this table. The significance of the cubic, octahedral or dodecahedral 'mode' as indicative of the lattice type no longer holds, for the predominant form may be {100}, {110}, or {111} in crystals based on a P lattice; {100}, {110} or {111} also in those based on an F lattice; and {100}, {110} or {211} in those based on an I lattice. The type of lattice must be deduced from consideration of the order of morphological importance shown by several of the predominant forms. In two aspects the form {211}—icositetrahedron or tristetrahedron—heads the list, a position never attained by it under the Bravais principle, but exemplified, for example, in the common habit of the garnets (Fig. 80), analcime, leucite and eulytine (Fig. 291); in two others it is second in importance, as illustrated, for example, by bixbyite. Notice also that of a given family of forms it by no means follows that the member with the

lowest index is most important; {311} may exceed {211} in importance
(gold, magnetite (Fig. 92)); {331} may be morphologically of greater
importance than its simpler relative {221}, as in the spinels and micro-
lite.   Certain aspects are characteristic of one space group only—consideration of the common habits of garnets places them unequivocally
in the group $Ia3d$—but others apply to two or more space groups
isomorphous with different point groups, and distinction between the
latter must be effected by some of the means which we have discussed
above (p. 151).

An interesting point arises in the space group $Pa3$, to which pyrite
can be uniquely assigned on account of its clear didodecahedral
morphology (class $m3$) and the observed order of importance {100},
{111}, {210}, ...   The pentagonal dodecahedron {210} appears third
on the list, in agreement with the frequent development of such a form.
The a glide planes, however, demand calculation of all $h\,k\,0$ planes
with h even, so that while {210} appears thus high in the list the reticular
area of the planes of its complementary form {120} must be calculated
from the index 240, and this form is relegated to the fourteenth place.
Only if both h and k are odd will complementary pentagonal dodecahedra $\{h\,k\,0\}$ and $\{k\,h\,0\}$ rank equal in importance; X-ray evidence
shows clearly that {210} is vastly more important than {120} as a growth
form, and Donnay and Harker have shown that so far as data are
available there is remarkable agreement between theoretical prediction
and the facts of observation concerning the development of pentagonal dodecahedra with higher indices.

Similar considerations arise when the generalised principle is applied
in other crystal systems, and these authors have tabulated 97 different
morphological aspects covering the 230 space groups.   This expansion
of the fourteen aspects arising from the Bravais Law proceeds from
similar considerations to those which we have employed in our discussion of the cubic system; pinacoids or prisms normal to screw axes
are decreased correspondingly in expected morphological importance,
whilst certain forms in every zone normal to glide planes likewise
acquire effectively larger reticular areas.   We may illustrate the expected resultant effect on crystal morphology by some simple calculations for a holosymmetric orthorhombic crystal based on a primitive
lattice.   Suppose that $a:b:c = 0\!\cdot\!9:1:0\!\cdot\!8$.   If the symmetry planes
parallel to 100 are reflection planes m, the order of increasing reticular
area in the zone [100] is

010, 001, 011, 021, 012, 031, 013, 032, 041, 023 ... .

If there are b glide planes parallel to 100, however, we must calculate only with indices $0\,k\,l$ in which k is even, and the appropriate order becomes

001, 010, 021, 011, 041, 023, 012, 043, 061, 031, ... ,

in which the specially noticeable feature is the predominance of the dome {021} over the simpler form {011}. The presence of c glide planes parallel to 100 gives the order

010, 001, 012, 011, 032, 021, 014, 052, 034, 031, ... ,

in which the predominant dome is now {012}. n glide planes would give the order

011, 010, 001, 031, 013, 021, 051, 012, 053, 015, ...

We may next calculate the order of increasing reticular area for $h\,k\,0$ planes, on the assumption that the planes of symmetry parallel to 001 are reflection planes m, and compare the relative importance of dome and prism forms. For any group Pm^*m we have:

Prisms	110		120		210		130		310, 230		320 ... .
Domes		011		021		012		031		013	

whilst any group Pb^*m gives

Prisms	110, 120		210		130,	310,	230,	320,	140	
Domes		021	011						041 ... .

This clearly implies a predominance of prisms over domes in the second case, so far as the morphological effect of these two kinds of plane of symmetry is concerned. We are reminded of the prominent development of domes {0 k l} in the carbonates of the aragonite group, for example (permutation $Pmcn$ of the space group $Pnma$ in the usual crystallographic setting) compared with the prismatic habit of such a substance as topaz (permutation $Pbnm$).

The predominance of particular forms is sometimes so closely related in this way to the characteristic elements of the underlying space group that it is possible to read the space group from an examination of a list of forms in order of relative importance, but we must first pay attention to a further important point. When choosing a parametral plane to establish the axial parameters for a given description of a crystal, we have been at some pains to point out that there is no one particular plane which must be selected; 'any plane, parallel to a crystal face, which is not parallel to any of the crystallographic axes' (p. 40) will serve, though we have often found that one choice may lead to a simpler set of indices than does an alternative choice. When consideration is given to the details of symmetry of the structural pattern, however, it is essential that the plane indexed 111 should in fact cut the crystallographic axes in the ratios of the lengths of the sides of the actual unit of pattern—the structure defines a set of 111 planes uniquely. The corresponding form can usually be selected unambiguously, for it will tend to be one of high morphological importance lying at the intersection of prominent zones. Thus for topaz Dana gives the axial ratios $a : b : c = 0.528 : 1 : 0.477$, but on this description $[1\bar{1}0]$, $[10\bar{2}]$ and $[01\bar{2}]$ are prominent zones, with $\{201\}$, $\{021\}$, $\{041\}$, $\{221\}$, $\{111\}$ and $\{223\}$ as forms of considerable morphological significance. If we double the vertical axis, by selecting Dana's plane 221 as the parametral plane, we simplify this list of forms to read $\{101\}$, $\{011\}$, $\{021\}$, $\{111\}$, $\{112\}$ and $\{113\}$ (Fig. 514), and we can confidently expect

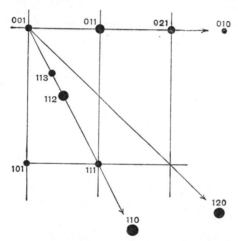

FIG. 514. A gnomonogram of topaz. The plane indexed 111 is Dana's 221. The sizes of the circles are proportional to the observed morphological importance of the corresponding forms.

on the arguments we are developing that the edges of the true unit of pattern will be in the ratios $a_0 : b_0 : c_0 = 0.528 : 1 : 0.954$. Again, in brookite, a study of the relative frequency of forms in terms of the mineralogists' axial ratios $a : b : c = 0.842 : 1 : 0.944$ gave the order

100, 110, 122, 001, 104, 021, 102, 112, 010, 210, 043, 111, 121 ... ,

and a glance at a projection shows that the plane here indexed 122 is the probable structural 111 plane (Fig. 515). In terms of this para-

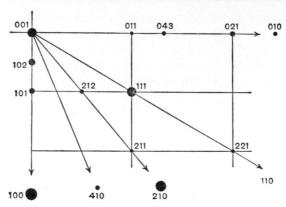

FIG. 515. A gnomonogram of brookite. The plane indexed 111 is the plane 122 of the customary mineralogical description. The sizes of the circles are proportional to the observed morphological importance of the corresponding forms

metral plane, the axial ratios become $a : b : c = 1\cdot684 : 1 : 0\cdot944$, and the observed morphological importance in the various groups of forms reads:

Bipyramids $\{h\,k\,l\}$ —111, 212, 211, 221, 234, 321 ... .
Domes $\{0\,k\,l\}$ —001, 021, 010, 043 ... .
  „ $\{h\,0\,l\}$ —100, 001, 102, 101 ... .
Prisms $\{h\,k\,0\}$—100, 210, 010, 410, 810 ... ,

in which we see clearly reflected the characteristics of the space group *Pbca*.

Examples could be multiplied indefinitely, but enough has been said to justify the conclusion that the detailed geometry of the structural arrangement, extending to the particular space group pattern on which the structure is built up, plays an important part in determining the *dominant habit of a crystal species* so far as the zonal development and size and frequency of occurrence of various forms are concerned. This statement in no way contradicts our earlier observations that the external environment during crystallisation plays an important part in determining the habit of *particular individuals*. Nor does it round off the study of crystal habit completely, for some flagrant anomalies remain outstanding, particularly amongst simple ionic crystals. Our discussion, in terms of reticular areas, has been conducted purely in terms of the *geometry* of the lattice: recent work on crystal habit attempts a more complete explanation by taking into account also the energetics of the crystal, involving a study of the actual positions of the structural units and the bond-strengths between them. We do at last, however, appear to be approaching a fuller understanding of some of the factors controlling the wide diversity of habit which perplexes, in particular, the student of minerals.

APPENDIX

NOTE ON SCHOENFLIES' NOTATION FOR POINT GROUPS AND SPACE GROUPS

Though the student is advised always to work in the elegant Hermann-Mauguin notation, which we have used, it will be necessary in more advanced work to consult older descriptions in which the notation devised by Schoenflies is employed. We therefore append a summary account of this notation.

Schoenflies described crystal symmetry in terms of a centre, reflection planes, rotation axes and rotary-reflection axes ('alternating' axes, p. 104). A single rotation axis of degree n is denoted by the symbol C (from *Cyclic*) and appropriate numerical suffix—C_n. Thus the symbols C_1, C_2, C_3, C_4 and C_6 are equivalent to our symbols 1, 2, 3, 4 and 6 respectively. The addition of a plane of symmetry normal to the axis (a horizontal plane) is symbolised $C_n{}^h$. (In earlier work the plane of symmetry in classes m and $2/m$ was conventionally set horizontal.) Hence

$C_1{}^h$ is the equivalent of m.
$C_2{}^h$,, ,, ,, $2/m$.
$C_3{}^h$,, ,, ,, $3/m$ (and hence of $\bar{6}$).
$C_4{}^h$,, ,, ,, $4/m$.
$C_6{}^h$,, ,, ,, $6/m$.

To denote the presence of a plane of symmetry through the axis (a vertical plane) Schoenflies wrote $C_n{}^v$.

$C_1{}^v$ is the equivalent of $C_1{}^h$, and is not used.
$C_2{}^v$,, ,, ,, $mm2$.
$C_3{}^v$,, ,, ,, $3m$.
$C_4{}^v$,, ,, ,, $4mm$.
$C_6{}^v$,, ,, ,, $6mm$.

An axis of rotary-reflection of degree n is written S_n (from *Sphenoidisch*).

S_1 is the equivalent of $C_1{}^h$, and is not used.
S_2 ,, ,, ,, $\bar{1}$.
S_3 ,, ,, ,, $C_3{}^h$, and is not used.
S_4 ,, ,, ,, $\bar{4}$.
S_6 ,, ,, ,, $\bar{3}$.

A rotation axis of degree n with n diad axes normal to it is written D_n (from *Diëdergruppe*).

D_1 is the equivalent of C_2, and is not used.

D_2 „ „ „ 222.
D_3 „ „ „ 32.
D_4 „ „ „ 422.
D_6 „ „ „ 622.

Adding to the groups D_n a horizontal plane we derive further groups $D_n{}^h$.

$D_2{}^h$ is the equivalent of *mmm*.
$D_3{}^h$ „ „ „ $\bar{6}m$ ($= 3/mm$).
$D_4{}^h$ „ „ „ 4/*mmm*.
$D_6{}^h$ „ „ „ 6/*mmm*.

The addition of a vertical plane to groups D_n instead of a horizontal plane would produce the same results if the vertical plane passes through one of the horizontal diad axes. To groups D_2 and D_3, however, we may add a diagonal vertical plane.

$D_2{}^d$ is the equivalent of $\bar{4}2m$.
$D_3{}^d$ „ „ „ $\bar{3}m$.

In the cubic system, the symbol T is used in tetrahedral symmetry groups (four triad axes with three diad axes) and the symbol O in octahedral groups (four triad axes and three tetrad axes).

T is the equivalent of 23.
T_h „ „ „ *m*3.
T_d „ „ „ $\bar{4}3m$.
O „ „ „ 432.
O_h „ „ „ *m*3*m*.

We have thus established the Schoenflies symbols for the 32 point groups, but for a few of these alternatives have been used. A centre of symmetry is denoted by the letter i. For even values of n, $C_n{}^i$ is, of course, equivalent to $C_n{}^h$ and the symbols are unnecessary; for odd values:

C_i denotes a centre alone ($= \bar{1}$), and is a synonym for S_2.

$C_3{}^i$ is equivalent to $\bar{3}$, and is a synonym for S_6.

A single plane is sometimes denoted by the letter s (from *Spiegelung*, reflection).

C_s is a synonym for $C_1{}^h$, and is perhaps to be preferred since crystallographers now conventionally set the plane vertical and not horizontal.

The equal significance of the three diad axes in groups 222 and *mmm* is sometimes expressed by a special symbol V (from *Vierergruppe*, leading some English authors to use the expression *Quadratic group* and the symbol Q).

$$V \text{ is a synonym for } D_2.$$
$$V_h \quad ,, \quad ,, \quad ,, \quad D_2{}^h.$$

The use of V_d as an alternative to $D_2{}^d$ seems less desirable, since the vertical axis in this group is an inversion tetrad axis and not merely a diad.

Triclinic		Monoclinic		Trigonal		Tetragonal		Hexagonal		Cubic	
1	C_1	2	C_2	3	C_3	4	C_4	6	C_6	23	T
$\bar{1}$	C_i, S_2	m	$C_s, C_1{}^h$	$\bar{3}$	$C_3{}^i, S_6$	$\bar{4}$	S_4	$\bar{6}$	$C_3{}^h$		
		$2/m$	$C_2{}^h$			$4/m$	$C_4{}^h$	$6/m$	$C_6{}^h$	$m3$	T_h
		Orthorhombic		$3m$	$C_3{}^v$	$4mm$	$C_4{}^v$	$6mm$	$C_6{}^v$		
		$mm2$	$C_2{}^v$	$\bar{3}m$	$D_3{}^d$	$\bar{4}2m$	$D_2{}^d, V_d$	$\bar{6}m2$	$D_3{}^h$	$\bar{4}3m$	T_d
		222	D_2, V	32	D_3	422	D_4	622	D_6	432	O
		mmm	$D_2{}^h, V_h$			$4/mmm$	$D_4{}^h$	$6/mmm$	$D_6{}^h$	$m3m$	O_h

The space groups isomorphous with each point group were numbered successively by Schoenflies in the order in which he derived them, the numbers being added to the symbol in the exponent position. Thus $C_2{}^1$, $C_2{}^2$ and $C_2{}^3$ are the space groups $P2$, $P2_1$ and $C2$ respectively of the point group C_2 ($=2$). Where the exponent position is occupied by a letter of the point group symbol, this letter is reduced to a subscript position in the space group symbols. Thus $D_{2h}{}^{14}$ is the space group $Pbcn$, the fourteenth of the groups isomorphous with the point group $D_2{}^h$ ($=mmm$). It is at this stage that the Hermann-Mauguin notation is so much to be preferred since (apart from any consideration for the compositor!) *each* space group symbol conveys all the essential information, whereas we must always have a catalogue at hand to follow an arbitrary serial numbering such as that of Schoenflies.

GENERAL INDEX

INDEX OF FORMULAE